Nanotechnology in Chemistry

Nanotechnology in Chemistry

Edited by:

Dr. Syed Ismail

ARCLER
PRESS

www.arclerpress.com

Nanotechnology in Chemistry

Dr. Syed Ismail

Arcler Press
4164 Lakeshore Road
Burlington, ON L7L 1A4
Canada
www.arclerpress.com
Email: orders@arclereducation.com

ISBN: 978-1-77956-415-3 (Hardcover)

ABOUT THE EDITOR

Dr. Syed Ismail has a total of 30 years working experience in teaching, research, extension and academic administration. His field of specialization is soil microbiology and biochemistry with special reference to micronutrient management in different crops. During his service carrier he worked in different capacities as Assistant Professor, Associate Professor, Professor, Head, Department of Soil Science and Agricultural Chemistry and worked as Principal Investigator /Co-PI in various ICAR Research Projects and one CFC-FAO-ICRISAT project. He served as Principal Investigator, ICAR- All India Network Project on Soil Biodiversity- Biofertilizers for eight years and also PI in ICAR-All India Coordinated Research Project on Long Term Fertilizer Experiments at Parbhani for three years. He has successfully guided 35 M.Sc. (Agri) and 10 Ph.D. students in the discipline of Soil Science and Agricultural Chemistry as a major advisor.

At present he is working as Associate Dean and Principal, College of Agriculture at Vasantrao Naik Marathwada Agricultural University. Dr. Ismail has many publications to his credit to mention: Books/book chapters:20, Papers published in per reviewed journals:119, Scientific reviews:16, Technical Bulletins :12, Papers presented in seminars/ symposia:90. He received ICAR sponsored Best Teacher's Award of VN Marathwada Agril. University, Parbhani for contribution in teaching, RS Malhotra Award of VNMKV for research contribution and state level Mridagandha Puraskar for the work carried out to enhance soil health through sustainable soil management practices. International Crop Research Institute for Semi-Arid Tropics (ICRISAT), Patancheru conferred special recognition to him for successful implementation of farmers' participatory project and recently received life time achievement award for outstanding contribution in the field of Soil Science and Agricultural Chemistry by Indian Society of Agriculture and Horticulture Research Development , Ludhiana.

TABLE OF CONTENTS

LIST OF FIGURES

LIST OF TABLES

PREFACE

Recent decades have seen rapid advancements in technology. The science of nanotechnology will play an essential role in this technological growth. Nanotechnology is expected to contribute significantly to global technological advancements, according to experts. The convergence of nanotechnology and chemistry has ushered in a new era of scientific discovery and technological innovation. Nanotechnology is the study of matter, materials, and technologies at the nanoscale, measuring 1 nm–100 nm in size. Nanotechnology is an interdisciplinary science that spans medicine, biology, chemistry, and engineering. Nanotechnology is gaining popularity and funding because of its potential for growth. Nanotechnology has numerous applications, including energy production, textiles, cosmetics, home products, medical equipment, antibacterial creams, medications, and diagnostics that impact everyday life.

This book comprises of eight-chapters. Chapter 1 lays the foundation by introducing the concept of nanotechnology and its historical development. In Chapter 2, the diverse methods for synthesizing and stabilizing nanoparticles are explored. The ability to manipulate materials at the nanoscale opens up endless possibilities, and this chapter provides a detailed look at the various techniques and their practical implications.

The third chapter is dedicated to the characterization techniques essential for studying nanomaterials as accurate characterization is crucial for understanding the properties and behavior of nanomaterials. Chapter 4 highlights the transformative impact of nanotechnology on heterogeneous catalysis.

Chapter 5 discusses nanotubes, particularly carbon nanotubes, and their role in energy storage. The unique properties of nanotubes make them ideal for applications in batteries, super capacitors, and fuel cells, promising significant advancements in energy technologies. Chapter 6 deals with nanocomposites for surface coatings, emphasizing their improved performance characteristics. Chapter 7 delves into the exciting field of nanomedicine, with a focus on carbon nanotubes. Chapter 8 addresses the social and ethical implications of nanotechnology. As with any transformative technology, it is essential to consider the broader impacts on society, including environmental, health, and safety concerns.

This book provides a thorough and accessible introduction to nanotechnology in chemistry, offering valuable insights for students and experienced researchers.

—Editor

CHAPTER 1

INTRODUCTION TO NANOTECHNOLOGY

LEARNING OBJECTIVES

At the end of this chapter, readers will be able to:

1. Understand and articulate the fundamental definition of nano and its significance
2. Comprehend why the properties of materials change at the nanoscale
3. Identify the different types of nanostructured materials
4. Learn about the history of nanomaterials
5. Comprehend the applications of nanotechnology in different fields
6. Understand the reliability issues of MEMS/NEMS

INTRODUCTORY EXAMPLE

In the realm of nanotechnology, practical applications are continually emerging, demonstrating the transformative potential of this field. One notable example is the work of IBM Research, a leader in advancing nanoscale science and technology. IBM has developed innovative approaches to create more efficient and powerful semiconductor devices using nanotechnology.

IBM's breakthrough in developing nanosheet transistors is a prime example of nanotechnology's impact. These nanosheet transistors, which are expected to succeed the current FinFET technology, offer enhanced performance by allowing greater control over electrical currents at the nanoscale. This advancement is critical for continuing the trend of Moore's Law, which predicts the doubling of transistors on a microchip approximately every two years, leading to more powerful and energy-efficient electronic devices.

Another significant application by IBM involves the use of carbon nanotubes in electronics. Carbon nanotubes possess exceptional electrical properties, including high conductivity and strength, making them ideal for next-generation electronic components. IBM's research in this area aims to develop faster and smaller transistors, which could revolutionize computing by significantly increasing processing speeds while reducing energy consumption.

These examples from IBM underscore the real-world impact of nanotechnology, illustrating how manipulating materials at the nanoscale can lead to significant advancements in electronics and other industries.

UNIT INTRODUCTION

Nanotechnology is an interdisciplinary domain that encompasses the manipulation and regulation of materials within the range of 1 to 100 nanometers. At this level of measurement, materials have distinct and frequently improved characteristics in comparison to their larger-scale equivalents. This chapter offers a thorough overview of nanotechnology, specifically addressing its definition, importance, and the unique characteristics that materials exhibit at the nanoscale. Nanomaterials exist in a multitude of forms, encompassing nanoparticles, nanowires, nanotubes, nanocomposites, and nanofilms. Every nanomaterial possesses distinct attributes and uses, which stimulate advancements in many sectors (Bhushan, 2016).

Gaining knowledge about the historical progression of nanotechnology is crucial, starting from Richard Feynman's influential lecture in 1959 that envisioned the ability to manipulate every atom, to the subsequent technological developments that have brought nanotechnology into existence today. The utilization of nanostructures by nature has served as a source of inspiration for contemporary nanotechnology. Biological systems, including viruses and cells, exhibit remarkable efficiency and functionality while operating at the nanoscale. These natural systems provide essential insights and guidance for the creation of synthetic nanomaterials (Asiyanbola & Soboyejo, 2008).

Having the capacity to see and control substances at the nanoscale is crucial for the progress of nanotechnology. Scanning Tunneling Microscopy (STM) and Atomic Force Microscopy (AFM) are methods that can be used to see and manipulate nanostructures. These techniques provide a way to have accurate control over the properties and interactions of these structures. Nanotechnology possesses extensive applications in various fields such as medical, electronic devices, power, and materials research (Silva, 2004). These applications provide the potential to transform industries through enhancements in effectiveness, reliability, and sustainability. Nevertheless, the dependability of micro-electro-mechanical systems (MEMS) and nano-electro-mechanical systems (NEMS) poses considerable difficulties. To ensure the endurance and functionality of these systems in real applications, it is necessary to address challenges such as wear out, stiction, and fatigue (Subramani et al., 2019).

This chapter establishes the groundwork for a more thorough investigation of nanotechnology, imparting the fundamental understanding required to comprehend the subsequent, more intricate examinations of the several facets and uses of this revolutionary discipline (Pokropivny et al., 2007).

1.1. DEFINITION OF NANO AND ITS SIGNIFICANCE

Learning Objectives

- Understand the meaning of the term nano.
- Comprehend the physical and chemical properties of materials such as gold at nanoscale.

The term 'nano' is linguistically based on the Greek word that means dwarf, and it is used to describe extremely small objects. A nanometer (nm) is equivalent to 0.000 000 001 m. Below are a few relevant analogies that accurately illustrate the nanometer scale (Laherto, 2010):

a. Dividing a one mm segment on a ruler into a million equal pieces results in a length of one nanometer.

b. The mean diameter of a human hair ranges from around 60,000 to 100,000 nanometers.

c. A newspaper has a thickness of approximately 100,000 nanometers.

d. One nanometer is equivalent to a line of 10 hydrogen atoms (Chellaram et al., 2014).

Nanotechnology and its related fields encompass a range of scientific disciplines that focus on the manipulation and utilization of materials, devices, and processes at the nanometer level. These fields have garnered significant attention from scientists and researchers, resulting in numerous advancements that are now integrated into everyday lives. Nanomaterials are substances that possess at least a single dimension within the nanometer range (Hullmann, 2007). Researchers have shown that materials at the nanoscale exhibit distinct physical and chemical characteristics in comparison to their larger equivalents. Gold (Au) exhibits distinct characteristics, including (i) a lustrous, malleable, and gleaming appearance; (ii) excellent thermal and electrical conductivity; (iii) chemical inertness, which contributes to its extensive use in the jewelry sector; and (iv) a high melting point of 1,064°C. Gold exhibits these qualities at macroscopic scales, beyond the nanometer range. When gold is reduced to the nanometer scale, its typical □ualities undergo significant modifications (Mohammed et al., 2024). For instance, gold that is nanometer-sized does not exhibit the characteristic gold color. At the nanometer scale, it exhibits semiconductor properties, functions as an effective catalyst, and its point of melting decreases to 300°C (as shown in Table 1.1). Materials in the nanoscale often display distinct optical, electronic, magnetic, and chemical characteristics that are not seen in their bigger counterparts. The distinct nanoscale characteristics of these □ualities have a significant influence on both technology and medicine. Nanomaterials have emerged as fundamental constituents of commonplace, contemporary technology, and significant contributors to technological progress in the 21st century (Forest et al., 2019).

Table 1.1. Physical and Chemical Properties of Gold at Macro-Scale and Nano-Scale

Gold (Regular Size)	**Small Pieces of Gold at Nanometer Scale**
Golden color	Not golden
Soft in nature	Is a semiconductor
Chemically it is inert	Effective against cancer cells
Melts at 1,064°C	Melts at 300°C

Source: Charles, Creative Commons License.

Did You Know?

Nanotechnology draws inspiration from nature's nanostructures, such as lotus leaves and butterfly wings, to design innovative materials with enhanced properties.

1.2. PROPERTIES OF MATERIALS AT NANO SCALE

Learning Objectives

- Comprehend the change in properties of materials at nano scale.
- Understand the □uantum confinement effect in nanomaterials

The variations in material characteristics when sized in the nanoscale are mostly attributed to two main factors:

1.2.1. Surface Area–Volume Ratio

The physical characteristics and kinetics of chemical reactions are closely correlated with the quantity of nanoparticles that are introduced to the surroundings. The reactivity is mostly influenced by the surface area of the substance. When materials are divided into smaller bits, the overall volume of the object remains the same, but the total surface area increases. As an illustration, a cube that measures 1 cm on each side has a volume of 1 cm^3 and a total surface area of 6 cm^2. By partitioning the cube into a million tiny cubes measuring 0.1 mm × 0.1 mm × 0.1 mm, the overall volume of the cube remains unchanged (Suttiponparnit et al., 2011). However, straightforward computations demonstrate that as a result of this partitioning, the combined surface area of the smaller cubes increases by a factor of 100 compared to the surface area of the original cube. When the cube is subdivided into smaller, imperceptible cubes measuring 1 nm × 1 nm × 1 nm, its exterior area experiences a 10-million-fold rise. Nanoscale materials exhibit bigger surface areas compared to equivalent amounts of their bigger-sized equivalents made from the same substance (Figure 1.1). This occurs due to the exposure of previously enclosed atoms and molecules, which then form surfaces. This phenomenon at the nanoscale is remarkable. Nanomaterials with exceptionally high surface-to-volume ratios have a more profound interaction with the surrounding environment (Ding et al., 2001).

1.2.2. Quantum Confinement Effect

The effect of quantum confinement occurs when the size of a material impacts its electrical and optical characteristics. When a semiconductor catches a photon with an energy level that is equal to or greater than its bandgap, it causes an electron to move from the valence band to the conduction band. As a result of this process, a vacancy is created in the valence band, leading to the formation of an exciton (Zhang et al., 2003). An exciton refers to the pairing of an electron and a hole, which are attracted to each other due to the electrostatic interactions caused by Coulomb forces. In accordance with the principles of quantum physics, the Bohr excitation radius represents the distance separating the hole and the excited electron. When the migration of electrons in a semiconductor material is not constrained or restricted, the material is typically bigger than the diameter of the Bohr exciton (Liu et al., 2020).

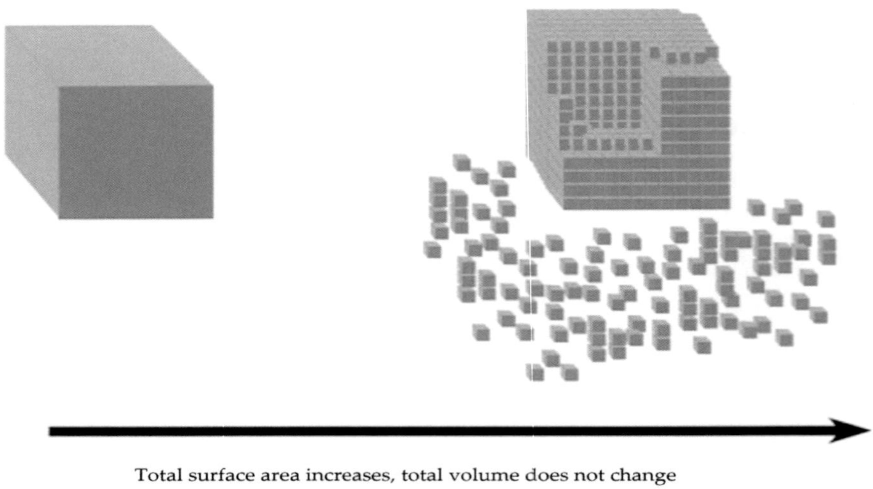

Figure 1.1. Illustration of point that total surface area increases while cutting cube to smaller cubes.

Source: Charles, Creative Commons License.

Conversely, when a material is reduced to the nanoscale, its size becomes less than the radius of the bulk Bohr exciton. Consequently, the electron motion is restricted and the electron holes are densely packed, leading to the quantum confinement of the electron-hole pairings. As a result, the bandgap expands, leading to the emergence of distinct energy levels. The higher bandgap leads to distinct optical and electrical characteristics. Consequently, materials that are nanoscale in size may exhibit colors that are not often observed in bigger elements of the same kind (Figure 1.2) (Takagahara & Takeda, 1992).

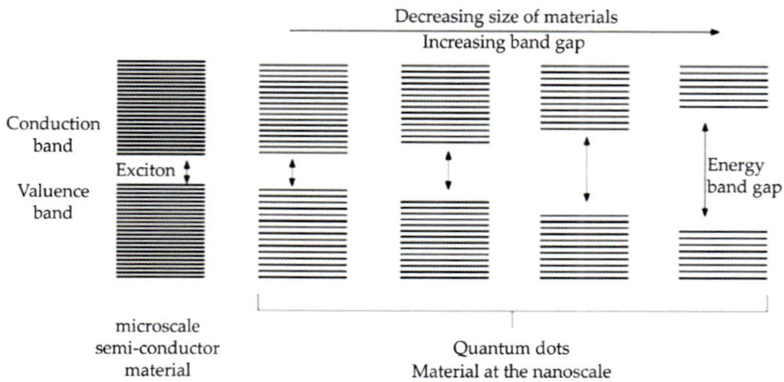

Figure 1.2. *Schematic illustration of quantum confinement effect.*

Source: Charles, Creative Commons License.

Remember

Properties of materials at the nanoscale are influenced by quantum mechanical effects, surface area to volume ratio, and dominance of surface forces.

1.3. TYPES OF NANOSTRUCTURED MATERIALS

Learning Objectives

- Understand the method to classify nanosized materials
- Comprehend different types of nanostructured materials

A prevalent approach for categorizing nanosized materials is by considering dimensions that fall beyond the nanoscale range. Thus, the four groups of nanomaterials are categorized as follows:

i. Zero-dimensional (0D) nanomaterials have dimensions ranging from 1 to 100 nm in each geometric dimension. Electron confinement happens in all three spatial dimensions: x, y, and z (Tiwari et al., 2012).

ii. One-dimensional (1D) structures have nanoscale dimensions in two dimensions, whereas one dimension is at the macro scale. Electron confinement takes place in a two-dimensional space, while in a one-dimensional space, electrons have unrestricted movement along it (Choi & Jang, 2010).

iii. Two-dimensional (2D) structures are composed of one dimension that is limited to the nanoscale and two additional dimensions that extend to the macro scale. Electrons are confined in one dimension, but they are unrestricted in their movement in the other two dimensions at a macroscale level (Babu et al., 2015).

iv. Three-dimensional (3D) architectures lack nanoscale dimensions. As a result, electron confinement in the nano band is absent, allowing electrons to travel unrestrictedly in the x, y, and z dimensions. Figure 1.3 depicts the categorization of nanostructured materials based on the number of dimensions at the nanoscale (Ren et al., 2018).

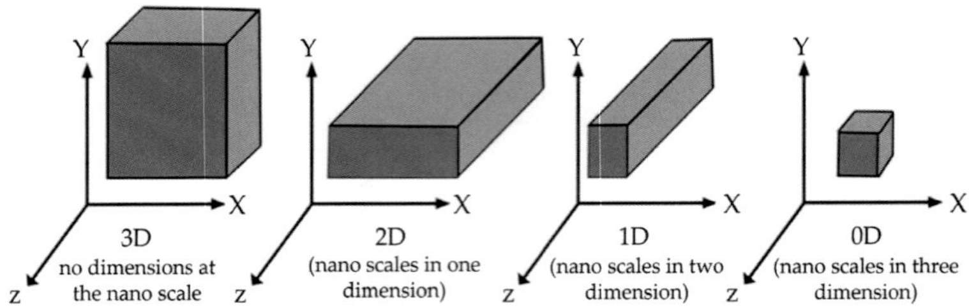

Figure 1.3. *Types of nanostructured materials based on number of nanoscale dimensions.*

Source: Charles, Creative Commons License.

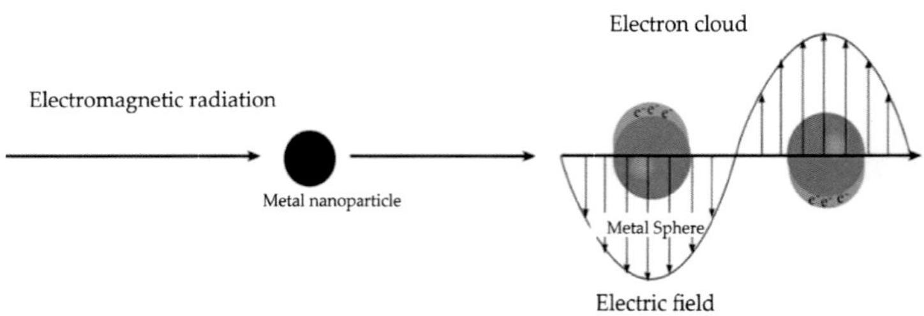

Figure 1.4. *Schematic illustration of localized surface plasmon resonance (LSPR) of metal nanoparticles.*

Source: Charles, Creative Commons License.

The LSPR phenomenon is influenced by the dimensions and morphology of metal nanoparticles. Gold nanoparticles have captivated scientists due to their capacity to exhibit a diverse range of hues, which arises from the synchronized motion of electrons on the exterior of the nanoparticles, as explained earlier in this chapter. The dimension of gold nanoparticles has a significant impact on optical absorption. As a result, gold nanoparticles can exhibit colors such as orange, red, purple, and sometimes blue. In order to create shiny finishes, painters in the 19th century took advantage of the wide variety of colorful metallic nanoparticles. They achieved this by either depositing the nanoparticles onto the surfaces of vessels or by creating stained glass. Quantum dots are minuscule particles composed of semiconductor materials with a nanostructured configuration. Quantum dots have a size range of 1 to 10 nm, which is smaller than that of nanoparticles (Gleiter, 2000).

Quantum dots exhibit a distinct range of colors due to the electron confinement phenomenon, which significantly defines their properties. Fullerene C60 is a molecular allotrope of carbon. Fullerene is a zero-dimensional nanoparticle composed of sp2 hybrid carbon atoms organized in five and six-membered rings. It has extensive applications in the fields of electronic devices, biological sciences, and pharmacology (Arabi, 2023).

Nanotubes are one-dimensional nanostructures with sizes ranging from 1 to 100 nm. Their lengths greatly exceed their diameter range. A nanowire is a wire-shaped solid substance having a diameter that falls within the nanoscale range. The aspect ratio refers to the ratio of the length and nanoscale diameter of a nanowire. Nanowires typically have aspect ratios exceeding 1,000, indicating that their length is roughly 1,000 times greater than their diameter (Koch, 2009). Various categories of nanowires encompass metal-based nanowires, semiconductor nanowires, and molecular nanowires. Nanowires consist of repetitive molecular units made of either inorganic or organic materials. Quantum confinement affects the diameters of two nanowires since their sizes are in the nanoscale range. The nanowire's length is not within the nanoscale range, hence the □uantum confinement effect is absent (Xu et al., 2019).

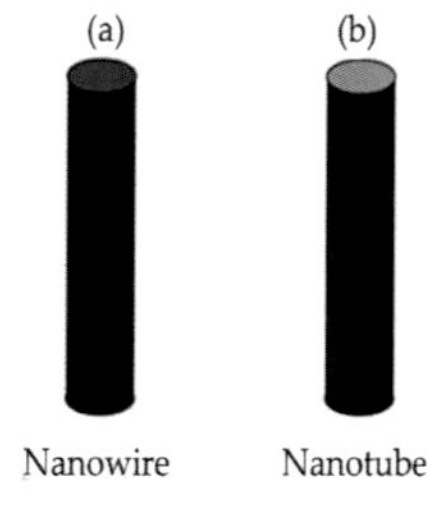

Figure 1.5. *Schematic description of (a) nanowire and (b) nanotube.*

Source: Charles, Creative Commons License.

Nanotubes are cylindrical nanowires that have a hollow structure. Single-walled carbon nanotubes (SWCNT) are the most widely recognized type of nanotubes. They are composed of graphene sheets that have been folded up. Graphene is a carbon allotrope. Graphene is a monolayer of graphite composed of carbon atoms arranged in hexagonal structures. Each carbon atom in graphene forms three covalent connections with other carbon atoms, resulting in one electron that is not involved in bonding and can move freely. This arrangement contributes to the high electrical conductivity of graphene (Rittner & Abraham, 1998) (Figures 1.5 and 1.6).

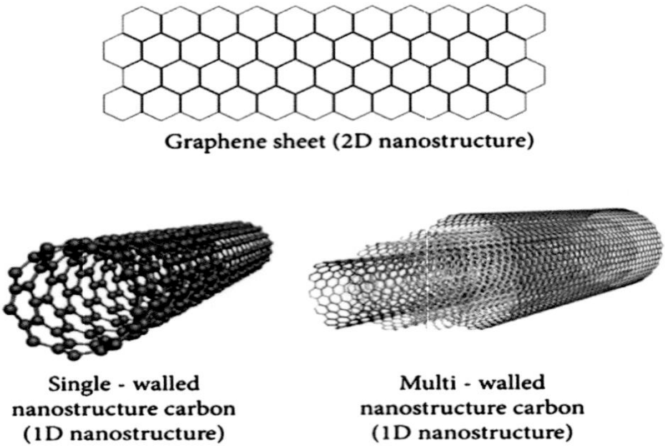

Figure 1.6. *Illustration of graphene sheet structure, single-walled carbon nanotube, and multiple-walled carbon nanotube.*

Source: Charles, Creative Commons License.

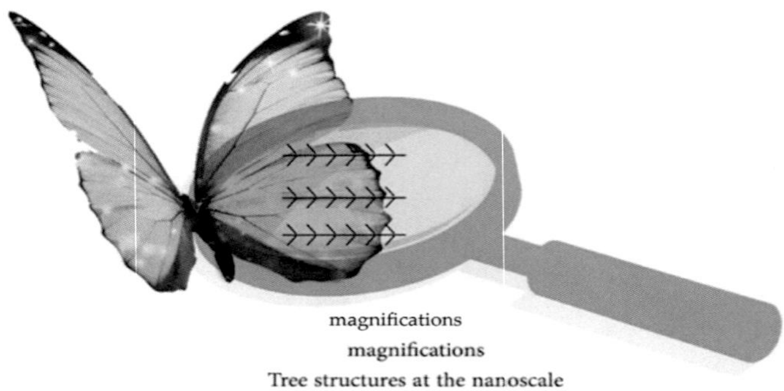

Figure 1.7. *Nanostructures of morpho butterfly wings.*

Source: Charles, Creative Commons License.

Nanowires, dispersed nanoparticles, and nanotubes can be used to create three-dimensional nanostructures. Various 3D nanostructures can be found in nature, such as the blue, iridescent-colored flaps of the morpho butterfly. The morpho butterfly's distinctive blue wings were found to be made up of complex 3D nanostructures, as depicted in Figure 1.7 (Leslie-Pelecky & Rieke, 1996).

1.4. HISTORY OF NANOMATERIALS

Learning Objectives

- Comprehend historical development of nanotechnology, from early conceptual ideas to modern advancements, and recognize key milestones and contributors in the field

Recent research has revealed that the utilization of nanostructured materials originated in ancient times, without any deliberate or conscious understanding of the concepts of nanoscience. Egyptian civilization, 5,000 years ago, was renowned for its production of a distinct and vibrant blue pigment known as Egyptian blue. During that era, the dye was commonly employed all through the region of the Middle East for painting, adorning monuments, embellishing tombs, and other similar applications. This dye exhibited a distinctive blue hue due to the presence of copper as one of its primary constituents (Barhoum et al., 2022). The hue of the object varied from a pale to a deep shade of blue, depending on its composition. Recent research has conclusively demonstrated that this distinct blue pigment is comprised of a blend of Cuprorivaite ($CaCuSi_4O_{10}$). To get the distinctive variety of blue tints, these compounds are subjected to high temperatures, often ranging from 800 to 900°C. Following the conclusion of the Roman era, the knowledge of the Egyptian blue dye, including its components and the process for its synthesis, was lost and remained in a state of uncertainty. The creation of colored stained glass emerged as a significant phenomenon in Europe during the late Bronze Age, specifically between the years 1,200 and 1,000 BC. An analysis of a red glass found in the Italian region of Frattesina di Rovigo has revealed that it is made up of copper nanoparticles. Additionally, copper nanoparticles have been discovered in red Celtic enamel dating back to 400–100 BC. The Lycurgus Cup, a renowned artifact, was crafted by Romans in the 4th or 5th century AD using gold and silver nanoparticles (Amen et al., 2021). The cup exhibits dichroism, displaying two distinct hues depending on the angle of incident light. When light enters the cup from the front, it gives the appearance of being green. When light is transmitted from the rear, the cup exhibits a red appearance. The dichroism phenomenon, characterized by the presence of two hues, is ascribed to the distinctive effects generated by the silver and gold nanoparticles in the cup. The Damascus Sword, crafted during the period spanning from the ninth to the 17th century, was distinguished by its exceptional amalgamation of pliability and robustness. The stiffness of swords, which are typically composed of steel, is determined by the quantities and ratios of carbon and iron in the steel (Kagdada et al., 2023).

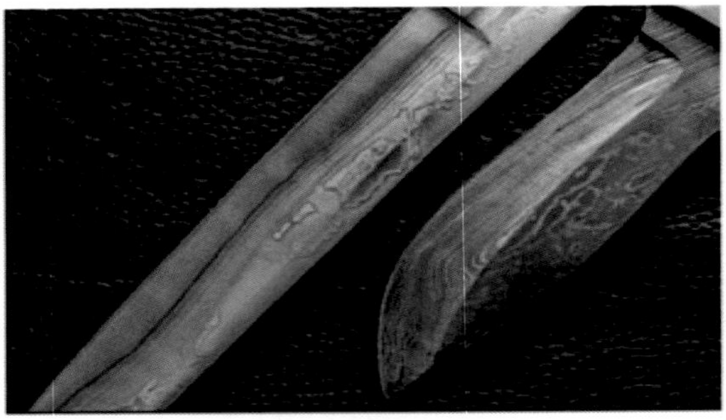

Figure 1.8. *Image of ancient Damascus sword.*

Source: Charles, Creative Commons License.

At the start of the 18th century, the Russians and French endeavored to replicate the robust and adaptable characteristics of the Damascus Sword. Nevertheless, their endeavors were futile, and they were incapable of unraveling the enigma of its creation. In 2006, scientists unveiled the manufacturing technique responsible for the distinctive characteristics of the sword. Figure 1.8 showcases the Damascus Sword, which incorporates carbon nanotubes, a very resilient and pliable material that is renowned for its durability (Baig et al., 2021). Metal nanoparticles exhibit unique hues that deviate from their typical appearance at larger scales, which can be attributed to the process of localized surface plasmon resonance (LSPR). During the Middle Ages in Europe, the process of coloring medieval stained glass involved painting window panes in churches using unique metal nanoparticle pigments. Subsequent research revealed that these colors were composed of liquids containing nanoparticles of silver, gold, and copper. The field of nanoscience and nanotechnologies was initially introduced to the scientific community in 1857 by the English scientist Michael Faraday. He pioneered the initial scientific methodology for generating the ruby hue of gold nanoparticles suspended in a water-based solution. In 1908, scientist Mie elucidated that the manifestation of diverse metallic hues at the nanoscale was attributable to the hypothesis of quantum size effects (Riaz et al., 2021). Silicon dioxide SiO_2 nanoparticles were initially manufactured and employed as additions for black carbon in rubber reinforcing in the year 1940. In 1959, Richard Feynman delivered a well-known talk at the American Physical Society meeting titled "There is Plenty of Room at the Bottom." Here he hypothesized, "What capabilities would scientists possess if they had the ability to manipulate matter at the atomic and molecular level and rearrange it according to their desires?" He also outlined a novel scientific discipline focused on manipulating small molecules and atoms to create materials and technologies with unique characteristics. This was the initial documented reference to a novel discipline thereafter referred to as nanotechnology. In 1974, the word 'nanotechnology' was used to refer to several techniques for creating minuscule particles. In 1976, Munir Nayfeh, a particle physicist of Palestinian and American descent, developed an advanced laser technology that enables the detection and precise measurement of individual atoms with unparalleled accuracy and control (Behera & Behera, 2022). He successfully detected a single atom from a large number and was able to determine its identification, marking the first time this has been achieved. His approach relied on using laser pulses to excite the atoms. In 1981, the Swiss scientists Gerd Ping and Henrik Rohr devised the Scanning Tunneling Microscope, building upon Munir Nayfeh's earlier work. This microscope enabled scientists to directly visualize and control individual atoms and molecules, facilitating the creation of nanoparticles, and marking a significant milestone in scientific history. Carbon nanotubes were developed by the Japanese scientist Sumio Lijima in 1991. The development of carbon nanotubes has facilitated the production of different nanostructures due to their exceptional mechanical and electrical capabilities (Sudha et al., 2018).

Practice Problem

Explain why quantum dots exhibit size-dependent optical properties?

Solutions to Practice Problem

Quantum dots are semiconductor nanoparticles that exhibit □uantum confinement effects due to their small size. When the size of the quantum dot approaches the exciton Bohr radius, the energy levels become discrete rather than continuous, similar to the behavior of atoms. This □uantum confinement effect means that the energy gap between the valence and conduction bands increases as the size of the quantum dot decreases.

Because the energy gap determines the wavelength (and thus the color) of light emitted or absorbed, smaller quantum dots will emit light at shorter wavelengths (bluer light), while larger quantum dots will emit light at longer wavelengths (redder light). This size-dependent optical property allows quantum dots to be precisely tuned for various applications in imaging, displays, and photovoltaics.

1.5. LESSONS FROM NATURE

Learning Objectives

- Understand how natural nanostructures inspire the design of synthetic nanomaterials, citing specific examples from biological systems.

Nanotechnology is a recently coined term, but, it does not represent a wholly novel discipline. The natural world encompasses several entities and phenomena that operate at a microscopic to nanoscopic level (Rist & Hartmann, 2018). Comprehending these functions can help in replicating and manufacturing nanodevices and nanomaterials. Several billion years in the past, molecules started to arrange themselves into intricate structures capable of sustaining life. Photosynthesis utilizes sun energy to sustain plant life. Plants include molecular ensembles consisting of light-collecting molecules, like chlorophyll, which are organized within cells at the nano to micrometer scales (Cai et al., 2022). These structures, known as chloroplasts, absorb light energy and transform it into chemical energy, which powers the metabolic processes within plant cells. Living organisms utilize chemical energy within their bodies. The flagella, a bacterial appendage, rotates at a speed exceeding 10,000 revolutions per minute (Arvidsson et al., 2013).

Within the field of tribology, certain biological systems possess surfaces that exhibit anti-adhesion properties. Firstly, numerous plant leaves, such as the lotus leaf, are coated with a hydrophobic cuticle (Nagai & Toyokuni, 2010). This cuticle is made up of a blend of sizable hydrocarbon molecules that possess a significant aversion to water. Furthermore, the surface exhibits a distinctive distribution of roughness. There have been reports indicating that the roughness of certain hydrophobic leaf surfaces might reduce wetness, resulting in a higher angle of contact of droplets of water on these surfaces (Zinger et al., 2020).

1.6. SEEING AT NANO SCALE

Learning Objectives

- Understand the mechanism of observing extremely small sample sizes
- Comprehend the working of transmission and scanning electron microscope

Observing and visualizing minuscule sample sizes that are imperceptible to the unaided eye is a necessity in numerous scientific disciplines such as biology, chemistry, and physics. Optical microscopy emerged at the end of the 16th century as a means of amplifying and examining minute things. The technique of an optical microscope relies on the utilization of visible rays and various types of lenses to amplify the size of a sample (Mo et al., 2009). The magnification techni□ue relies on illuminating the sample directly using a two-lens setup. Subsequently, the optical microscope has the capability to magnify objects to an aggregate of 2000 times. This feature enables the user to observe germs and cells within the micrometer size range. Nevertheless, the light microscope's restricted magnification capability proved inade□uate for researchers engaged in studies that necessitated observation and analysis of items less than the microscale (Jiang et al., 2022).

Electron microscopy was conceived in 1931 by German engineers named Max Knoll and Ernst Ruska. The primary modification made to microscopic technology with the electron microscope was the substitution of visible light with electron beams. J. J. Thomson discovered the wave characteristics of electrons in a cathode ray study conducted in 1897. Einstein initially postulated that light is composed of discrete energy units called photons, which led to the recognition of the wave-particle duality of light. Louis de Broglie expanded the notion of wave-particle duality to encompass electrons. He put out the notion that electrons possess characteristics of both particles and waves and suggested an equation that represents the wavelength of a beam of electrons as (Ruivenkamp & Rip, 2010.:

$$\lambda = h/p$$

The symbol λ represents the de Broglie wavelength of an electron, h represents Planck's constant, and p represents the momentum of the electron. When contrasting the wavelength of electrons to that of visible light beams employed in optical microscopy instruments, the electron's wavelength is significantly shorter (Tselev et al., 2016). The standard wavelength of sunlight is approximately 5,000 angstroms, whereas the wavelength of electron beams only extends to 1 angstrom. The identification of the wave nature of electrons is regarded as a significant breakthrough in the development of electron wave-based microscopes, as opposed to those utilizing visible light beams. As a result, there was a significant increase in magnification capability, thanks to the extremely small wavelength of electron waves (Viljoen et al., 2021). Electron microscopes have the capability to achieve a magnification power of up to 2.5 million times, allowing for the observation of samples at the molecular or atomic level. Electron microscopes revolutionized the field of nanoscale particle analysis by providing a crucial and practical tool. Here are many electron microscope types commonly utilized for studying nanoscale materials and structures (Scholes & Rumbles, 2006).

1.6.1. Transmission Electron Microscope

The adoption of electron beams rather than visible light rays in magnification equipment has significantly influenced the advancement of nanotechnology. A transmission electron microscope (TEM) is a significant magnification instrument utilized in this particular domain (Inkson, 2016). An electron microscope is a type of microscope that achieves magnification by utilizing electron beams that pass through a material. Subsequently, a picture can be acquired either on a fluorescent screen or through the utilization of a charged-coupled device (CCD) camera. The high precision magnification capability of a Transmission Electron Microscope (TEM) is attributed to the tiny wavelength of electrons. Consequently, this gadget became a crucial analytical instrument employed in nanomaterial research (Morgan & Gilman, 2003).

1.6.2. Scanning Electron Microscope

A scanning electron microscope (SEM) is an instrument capable of capturing three-dimensional images to analyze the morphology and shape of objects. The scanning electron microscope (SEM) directs a beam of electrons onto a sample, allowing for a comprehensive three-dimensional analysis of the sample's surface by mapping intricate details such as cavities, scuffs, and fissures (Jinschek, 2014). A magnetic field is employed to manipulate the trajectory of the beam of electrons across the sample. The electron beam possesses significant kinetic energy, which is subsequently dissipated as a diverse range of signals resulting from interactions between the electrons and the solid sample, causing the electrons to decelerate (Harmer & Columba, 2010).

1.7. APPLICATIONS OF NANOTECHNOLOGY

Learning Objectives

- Identify the wide range of applications of nanotechnology in fields such as medicine, electronics, energy, and materials science.

Advancements in science and technology are enabling the production of micro and nanodevices for a wide range of commercial, consumer, and biomedical uses. Various MEMS equipment have been created, with some being commercially utilized. Diverse sensors are employed in commercial, consumer, and medical fields. Microinstruments as well as other industrial uses, such as micromirror arrays, utilize a range of different microstructures or microcomponents. Two of the most significant industrial applications that have a significant impact are accelerometers, with over 85 million units sold in 2002, and digital micromirror devices, with sales of around $400 million in 2001. Since 1991, integrated capacitive-type silicon acceleration sensors have been employed for airbag deployment in automobiles (Haleem et al., 2023). In 2001, the accelerometer technology sector was valued at approximately 1 billion dollars per year. Analog Devices was the leading company in this industry, next to Motorola and Bosch. Texas Instruments introduced commercial digital lighting processing (DLP) technology in 1996. These devices utilize digital micromirror devices (DMD) and are used for digital projector displays in compact and home cinema projectors, and also in table-top and projection TVs. Over a million projectors had been traded before 2002. Additional significant industrial applications encompass pressure gages, inkjet printing heads, and optic relays. NovaSensor introduced silicon-based piezo-resistive pressure gages for measuring manifold absolute pressure in engines in 1991. By 2002, its yearly sales reached over 25 million units. In 2002, the yearly sales of inkjet printing heads including microscale functional components amounted to around 400 million units. Motorola introduced capacitive pressure sensors specifically designed for tire pressure measurements (Zdrojewicz et al., 2015). Additional uses of MEMS devices encompass chemical sensors, gas sensors, infrared detectors, and focal plane arrays employed in earth observations, space research, and anti-missile defense. They are also utilized in pico-satellites for space applications, as well as several pneumatic, hydraulic, and other consumer items. MEMS devices are now being researched for use in magnetic storage systems, specifically for the development of magnetic disk drives with super-compact dimensions and ultra-high recording density. Multiple head/suspension microdevices with integrated components have been created specifically for contact recording purposes. Micro-actuators with high bandwidth and servo control have been created for applications requiring ultrahigh track density. These micro-actuators function as the precise position control component of a two-stage servo system, working in conjunction with a traditional actuator. Furthermore, little motors and actuators, measuring in millimeters, have been successfully manufactured for the purpose of tip-based recording methods (Ali Mansoori et al., 2008).

BIOMEMS are being utilized more frequently in both commercial and defense sectors. BIOMEMS finds applications in various fields, such as the use of biofluidic chips (also referred to as microfluidic chips, bioflips, or bio chips) for conducting chemical and biological analyzes. These biosensors are particularly useful in medical diagnostics, enabling the detection of proteins, RNA, DNA, and cells, arterial blood pressure assays, and toxins. Additionally, BIOMEMS are utilized in implantable devices for the delivery of pharmaceutical drugs (Seil & Webster, 2012). Biosensors, often known as lab-on-a-chip devices, combine sample manipulation, segregation, identification, and data analysis into a single platform. Biosensors are specifically engineered to identify either a singular or a group of (bio)chemicals, or to possess comprehensive analytical capabilities for a wide variety of (bio)chemical species, which are referred to as micro total analysis systems (μTAS). The chips utilize microfluidics technology and entail the precise control of minuscule volumes of fluids in tiny channels employing microvalves for a range of analytical purposes. An external pump is typically used to introduce the test fluid into the chip for analysis. Certain chips have been specifically engineered to include an inbuilt micropump that operates using an electrostatically operated diaphragm. Nova-Sensor created portable blood-pressure sensor chips made of silicon in the early 1990s to monitor blood pressure. These chips were widely adopted, with over 20 million units sold in 2002 (Rashidi & Khosravi-Darani, 2011).

Following the tragic events of September 11, 2001, there has been a heightened focus on the threat of biological and chemical warfare. As a result, handheld devices equipped with sensors capable of detecting germs, chemical-based or nerve agents, mustard agents, and chemical precursors have been developed. These devices are intended to safeguard subway systems, airports, the water supply, and the general population (Smith et al., 2013).

Additional BIOMEMS uses encompass minimally invasive surgical procedures, such as endoscopy, laser beam angioplasty, and microsurgery. It is also possible to create implantable synthetic organs. Micro-instruments manipulators are employed for the purpose of manipulating, positioning, examining, creating patterns on, and analyzing objects and characteristics that are at the nanoscale. Miniaturized analytical equipment comprises gas chromatography and mass spectrometry (MS). Additional equipment includes the micro-STM, with STM being an abbreviation for scanning tunneling microscope. Nanoelectromechanical systems (NEMS) encompass a wide range of nanocomponents, nanodevices, nano-systems, and nanomaterials (Surendiran et al., 2009).

Did You Know?

Nanotechnology is revolutionizing medicine by enabling targeted drug delivery, precise imaging, and regenerative therapies for conditions like cancer and neurological disorders

1.8. RELIABILITY ISSUES OF MEMS/NEMS

Learning Objectives

- Understand the reliability challenges associated with MEMS and NEMS, including wear, stiction, and fatigue, and the importance of addressing these issues for practical implementation.

There is a growing demand for a multidisciplinary and system-oriented strategy for producing micro/nanodevices that exhibit consistent functionality. Only by combining ideas from several fields and facilitating the exchange of information and personnel between research organizations can this be accomplished. Key failure mechanisms that must be addressed to enhance the dependability of MEMS/NEMS devices include bonding, abrasion, wear out, fractures, fatigue, and impurity. Surface micro/nano machined structures typically consist of polished and chemically reactive surfaces. MEMS/NEMS devices are highly susceptible to stiction, which is characterized by high static friction, due to their huge area-to-volume ratio (Isono, 2004). Fracture arises when the applied force on a microdevice surpasses the material's strength. Fracture is a significant issue regarding reliability, especially for the fragile materials utilized in the fabrication of these parts since it might promptly or gradually result in disastrous failures. In addition, the fracture of microstructures can result in the formation of debris, which can then trigger subsequent failure processes. Repeated loading over an extended period of time can create fatigue in materials that are not easily broken, which in turn can result in the device breaking and fracturing. Generally, this style of failure is quite straightforward to detect and easy to anticipate. Nevertheless, the lack of knowledge of the substance characteristics of thin films typically leads to inaccuracies in fatigue forecasts (Young et al., 2010).

A significant number of MEMS/NEMS devices function in close proximity to their maximum thermal dissipation capacity. They might come across areas of high temperature that can lead to malfunctions, especially in vulnerable components like diaphragms or cantilevers. Thermal fluctuations can lead to thermal strain and relaxation, which in turn can induce material breakdown and stress in cantilevers. Significant fluctuations in temperature, such as those encountered in outer space, can cause bimetallic beams to deform because of differences in their coefficient of thermal expansion. The issue of wrapping has been a significant challenge (van Spengen et al., 2017). The contamination, likely occurring during packaging and storage, can significantly impact the dependability of MEMS/NEMS devices. For instance, the deposition of a particle of dust on an electrode of a comb drive might lead to severe and irreversible damage. The absence of MEMS/NEMS fabrication standards poses a challenge in transferring fabrication procedures between foundries.

Undoubtedly, doing research on identifying and controlling active failure modes that impact this emerging and promising technology is crucial for ensuring the high dependability of MEMS/NEMS devices and plays a decisive role in their successful real-world implementation (Feng et al., 2017).

Specimen size has been observed to have an impact on mechanical characteristics. The study of mechanical properties of nanometer-scaled components is conducted to aid in the design of dependable systems, as strong mechanical qualities are crucial in such applications. The qualities of interest include Young's modulus, bend strength, hardness, fracture resilience, and fatigue life. The study employs finite element modeling to investigate the impact of surface smoothness and scratching on stress levels in nanostructures. When nanostructures are thinner than a critical physical length scale, the traditional theory may become invalid, and novel phenomena may arise. Molecular mechanics is employed to model the dynamics of a nanostructure (Achanta & Celis, 2015).

SUMMARY

- This chapter provided a comprehensive introduction to nanotechnology, with a focus on its definition, importance, and the distinctive characteristics of materials on the nanoscale. Nanotechnology is the manipulation of matter at a scale between 1 and 100 nanometers. It utilizes the unique quantum mechanical phenomena and the large area-to-volume ratios of nanomaterials to improve chemical, physical, and biological characteristics.
- The chapter provided an overview of many categories of nanomaterials, such as nanoparticles, nanowires, nanotubes, nanocomposites, and nanofilms, each possessing distinct properties and serving specialized purposes.
- The text explores the historical progression of nanotechnology, starting with Richard Feynman's visionary ideas and leading up to the present-day developments in microscopy methods that allow for the observation and control of particular atoms and molecules. The utilization of efficient nanostructures observed in biological systems was emphasized as a source of inspiration for the design of synthetic nanomaterials. The wide-ranging uses of nanotechnology in the disciplines of medical, electronic devices, power, and materials research were explored, highlighting its capacity to profoundly transform these areas.
- The chapter concluded by discussing the dependability concerns associated with MEMS and NEMS. It highlighted the key obstacles, including wear, stiction, and fatigue, that need to be resolved for their successful practical application.

MULTIPLE CHOICE QUESTIONS

1. What is the size range that defines nanotechnology?

a. 1 to 100 nanometers

b. 10 to 500 nanometers

c. 100 to 1000 nanometers

d. 1 to 1000 micrometers

2. Which of the following properties is significantly enhanced due to the high surface area to volume ratio at the nanoscale?

a. Thermal conductivity

b. Electrical insulation

c. Chemical reactivity

d. Magnetic permeability

3. What type of nanomaterial consists of cylindrical structures with a high aspect ratio?

a. Nanoparticles

b. Nanotubes

c. Nanofilms

d. Nanocomposites

4. Who is credited with the lecture titled "There's Plenty of Room at the Bottom," which envisioned the potential of manipulating individual atoms?

a. Richard Feynman

b. Norio Taniguchi

c. Eric Drexler

d. James Tour

5. Which biological structure is known for its nanoscale photonic structures that manipulate light?

a. Butterfly wings

b. Spider silk

c. Human hair

d. Bird feathers

6. What technique uses a sharp probe to scan the surface of a material at the nanoscale, providing topographical data?

a. Transmission Electron Microscopy (TEM)

b. Scanning Tunneling Microscopy (STM)

c. X-ray Diffraction (XRD)

d. Nuclear Magnetic Resonance (NMR)

7. What is a primary challenge associated with the reliability of MEMS/NEMS?

a. Excessive thermal expansion

b. Stiction

c. Low electrical resistance

d. High optical transparency

hydrogen generation. *Physical Chemistry Chemical Physics*, *17*(5), 2960–2986.

8. Baig, N., Kammakakam, I., & Falath, W. (2021). Nanomaterials: A review of synthesis methods, properties, recent progress, and challenges. *Materials Advances*, *2*(6), 1821–1871.
9. Barhoum, A., García-Betancourt, M. L., Jeevanandam, J., Hussien, E. A., Mekkawy, S. A., Mostafa, M., & Bechelany, M. (2022). Review on natural, incidental, bioinspired, and engineered nanomaterials: History, definitions, classifications, synthesis, properties, market, toxicities, risks, and regulations. *Nanomaterials*, *12*(2), 177.
10. Behera, A., & Behera, A. (2022). Nanomaterials. *Advanced Materials: An Introduction to Modern Materials Science*, *5*(3), 77–125.
11. Bhushan, B. (2016). Introduction to nanotechnology: History, status, and importance of nanoscience and nanotechnology education. *Global Perspectives of Nanoscience and Engineering Education*, *7*(3), 1–31.
12. Cai, J., Chen, H., Li, Y., & Akbarzadeh, A. (2022). Lessons from nature for carbon-based nanoarchitected metamaterials. *Small Science*, *2*(12), 2200039.
13. Chellaram, C., Murugaboopathi, G., John, A. A., Sivakumar, R., Ganesan, S., Krithika, S., & Priya, G. (2014). Significance of nanotechnology in food industry. *APCBEE Procedia*, *8*, 109–113.
14. Choi, K. J., & Jang, H. W. (2010). One-dimensional oxide nanostructures as gas-sensing materials: Review and issues. *Sensors*, *10*(4), 4083–4099.
15. Ding, Y., Zhang, G., Wu, H., Hai, B., Wang, L., & Qian, Y. (2001). Nanoscale magnesium hydroxide and magnesium oxide powders: Control over size, shape, and structure via hydrothermal synthesis. *Chemistry of Materials*, *13*(2), 435–440.
16. Feng, P. X. L., Young, D. J., & Zorman, C. A. (2017). MEMS/NEMS devices and applications. In *Springer Handbook of Nanotechnology* (Vol. 5, pp. 395–429).
17. Forest, V., Pourchez, J., Guibert, C., Bitounis, D., Leclerc, L., Sarry, G., & Vergnon, J. M. (2019). Nano to micron-sized particle detection in patients' lungs and its pathological significance. *Environmental Science: Nano*, *6*(5), 1343–1350.
18. Gleiter, H. (2000). Nanostructured materials: Basic concepts and microstructure. *Acta Materialia*, *48*(1), 1–29.
19. Haleem, A., Javaid, M., Singh, R. P., Rab, S., & Suman, R. (2023). Applications of nanotechnology in medical field: A brief review. *Global Health Journal*, *7*(2), 70–77.

20. Harmer, A. J., & Columba, L. (2010). Engaging middle school students in nanoscale science, nanotechnology, and electron microscopy. *Journal of Nano Education*, *2*(1–2), 91–101.
21. Hullmann, A. (2007). Measuring and assessing the development of nanotechnology. *Scientometrics*, *70*(3), 739–758.
22. Inkson, B. J. (2016). Scanning electron microscopy (SEM) and transmission electron microscopy (TEM) for materials characterization. In *Materials Characterization Using Nondestructive Evaluation (NDE) Methods* (Vol. *2*, pp. 17–43). Woodhead Publishing.
23. Isono, Y. (2004). Micro/nano materials testing for reliable design of MEMS/NEMS. In *Micro-Nanomechatronics and Human Science, 2004 and The Fourth Symposium Micro-Nanomechatronics for Information-Based Society, 2004*. IEEE (Vol. *2*, pp. 33–38).
24. Jiang, Y., Sun, J., Cao, X., Liu, H., Xiong, C., & Nie, Z. (2022). Laser desorption/ionization mass spectrometry imaging: A new tool to see through nanoscale particles in biological systems. *Chemistry–A European Journal*, *28*(13), 202103710.
25. Jinschek, J. R. (2014). Advances in the environmental transmission electron microscope (ETEM) for nanoscale in situ studies of gas–Solid interactions. *Chemical Communications*, *50*(21), 2696–2706.
26. Kagdada, H. L., Bhojani, A. K., & Singh, D. K. (2023). An overview of nanomaterials: History, fundamentals, and applications. *Nanomaterials: Advances and Applications*, *1*(2), 1–26.
27. Koch, C. C. (2009). Nanostructured materials: An overview. *Bulk Nanostructured Materials*, *5*(3), 1–20.
28. Laherto, A. (2010). An analysis of the educational significance of nanoscience and nanotechnology in scientific and technological literacy. *Science Education International*, *21*(3), 160–175.
29. Leslie-Pelecky, D. L., & Rieke, R. D. (1996). Magnetic properties of nanostructured materials. *Chemistry of Materials*, *8*(8), 1770–1783.
30. Liu, J., Nie, Y., Xue, W., Wu, L., Jin, H., Jin, G., & Fu, C. (2020). Size effects on structural and optical properties of tin oxide quantum dots with enhanced □uantum confinement. *Journal of Materials Research and Technology*, *9*(4), 8020–8028.
31. Mo, Y., Turner, K. T., & Szlufarska, I. (2009). Friction laws at the nanoscale. *Nature*, *457*(7233), 1116–1119.
32. Mohammed, N., Nawar, S. H., Etawy, M. S., Nassar, G. E., & Hassabo, A. G. (2024). Nanotechnology and its applications in industry and product

design. *Journal of Textiles, Coloration and Polymer Science*, *21*(2), 273–284.

33. Morgan, A. B., & Gilman, J. W. (2003). Characterization of polymer-layered silicate (clay) nanocomposites by transmission electron microscopy and X-ray diffraction: A comparative study. *Journal of Applied Polymer Science*, *87*(8), 1329–1338.
34. Nagai, H., & Toyokuni, S. (2010). Biopersistent fiber-induced inflammation and carcinogenesis: Lessons learned from asbestos toward safety of fibrous nanomaterials. *Archives of Biochemistry and Biophysics*, *502*(1), 1–7.
35. Pokropivny, V., Lohmus, R., Hussainova, I., Pokropivny, A., & Vlassov, S. (2007). Introduction to nanomaterials and nanotechnology. Tartu: Tartu University Press. (Vol. *3*, pp. 45–100).
36. Rashidi, L., & Khosravi-Darani, K. (2011). The applications of nanotechnology in food industry. *Critical Reviews in Food Science and Nutrition*, *51*(8), 723–730.
37. Ren, G., Liu, Y., Wang, W., Wang, M., Zhang, Z., Liang, Y., & Shen, J. (2018). Facile synthesis of highly active three-dimensional urchin-like Pd@ PtNi nanostructures for improved methanol and ethanol electrochemical oxidation. *ACS Applied Nano Materials*, *1*(7), 3226–3235.
38. Riaz, U., Mehmood, T., Iqbal, S., Asad, M., Iqbal, R., Nisar, U., & Masood Akhtar, M. (2021). Historical background, development and preparation of nanomaterials. *Nanotechnology: Trends and Future Applications*, *4*(3), 1–13.
39. Rist, S., & Hartmann, N. B. (2018). Aquatic ecotoxicity of microplastics and nanoplastics: Lessons learned from engineered nanomaterials. In *Freshwater Microplastics: Emerging Environmental Contaminants?* (Vol. *8*(7), pp. 25–49).
40. Rittner, M. N., & Abraham, T. (1998). Nanostructured materials: An overview and commercial analysis. *JOM*, *50*(1), 36.
41. Ruivenkamp, M., & Rip, A. (2010). Visualizing the invisible nanoscale: Study of visualization practices in nanotechnology community of practice. *Science & Technology Studies*, *23*(1), 3–36.
42. Scholes, G. D., & Rumbles, G. (2006). Excitons in nanoscale systems. *Nature Materials*, *5*(9), 683–696.
43. Seil, J. T., & Webster, T. J. (2012). Antimicrobial applications of nanotechnology: Methods and literature. *International Journal of Nanomedicine*, *85*, 2767–2781.

44. Silva, G. A. (2004). Introduction to nanotechnology and its applications to medicine. *Surgical Neurology*, *61*(3), 216–220.

45. Smith, D. M., Simon, J. K., & Baker Jr, J. R. (2013). Applications of nanotechnology for immunology. *Nature Reviews Immunology*, *13*(8), 592–605.

46. Subramani, K., Elhissi, A., Subbiah, U., & Ahmed, W. (2019). Introduction to nanotechnology. In *Nanobiomaterials in Clinical Dentistry* (Vol. *2*, pp. 3–18). Elsevier.

47. Sudha, P. N., Sangeetha, K., Vijayalakshmi, K., & Barhoum, A. (2018). Nanomaterials history, classification, uni□ue properties, production and market. In *Emerging Applications of Nanoparticles and Architecture Nanostructures* (Vol. *1*, pp. 341–384). Elsevier.

48. Surendiran, A., Sandhiya, S., Pradhan, S. C., & Adithan, C. (2009). Novel applications of nanotechnology in medicine. *Indian Journal of Medical Research*, *130*(6), 689–701.

49. Suttiponparnit, K., Jiang, J., Sahu, M., Suvachittanont, S., Charinpanitkul, T., & Biswas, P. (2011). Role of surface area, primary particle size, and crystal phase on titanium dioxide nanoparticle dispersion properties. *Nanoscale Research Letters*, *6*, 1–8.

50. Takagahara, T., & Takeda, K. (1992). Theory of the □uantum confinement effect on excitons in quantum dots of indirect-gap materials. *Physical Review B*, *46*(23), 15578.

51. Tiwari, J. N., Tiwari, R. N., & Kim, K. S. (2012). Zero-dimensional, one-dimensional, two-dimensional and three-dimensional nanostructured materials for advanced electrochemical energy devices. *Progress in Materials Science*, *57*(4), 724–803.

52. Tselev, A., Velmurugan, J., Ievlev, A. V., Kalinin, S. V., & Kolmakov, A. (2016). Seeing through walls at the nanoscale: Microwave microscopy of enclosed objects and processes in liquids. *ACS Nano*, *10*(3), 3562–3570.

53. van Spengen, W. M., Modliński, R., Puers, R., & Jourdain, A. (2017). Failure mechanisms in MEMS/NEMS devices. In *Springer Handbook of Nanotechnology* (Vol. *2*, pp. 1437–1457).

54. Viljoen, A., Räth, E., Mckinney, J. D., Fantner, G. E., & Dufrêne, Y. F. (2021). Seeing and touching the mycomembrane at the nanoscale. *Journal of Bacteriology*, *203*(10), 10–1128.

55. Xu, C., Anusuyadevi, P. R., Aymonier, C., Luque, R., & Marre, S. (2019). Nanostructured materials for photocatalysis. *Chemical Society Reviews*, *48*(14), 3868–3902.

56. Young, D. J., Zorman, C. A., & Mehregany, M. (2010). MEMS/NEMS devices and applications. In *Springer Handbook of Nanotechnology* (Vol. *8*, pp. 359–387).

57. Zdrojewicz, Z., Waracki, M., Bugaj, B., Pypno, D., & Cabała, K. (2015). Medical applications of nanotechnology. *Advances in Hygiene and Experimental Medicine*, *69*, 1196–1204.

58. Zhang, L. Z., Sun, W., & Cheng, P. (2003). Spectroscopic and theoretical studies of □uantum and electronic confinement effects in nanostructured materials. *Molecules*, *8*(1), 207–222.

59. Zinger, A., Brozovich, A., Pasto, A., Sushnitha, M., Martinez, J. O., Evangelopoulos, M., & Taraballi, F. (2020). Bioinspired extracellular vesicles: Lessons learned from nature for biomedicine and bioengineering. *Nanomaterials*, *10*(11), 2172.

CHAPTER 2

SYNTHESIS AND STABILIZATION OF NANOPARTICLES

LEARNING OBJECTIVES

At the end of this chapter, readers will be able to:

1. Comprehend chemical reduction method for synthesis of nanoparticles
2. Understand the photochemical and radiation-chemical reduction techniques
3. Gain insight into the cryochemical synthesis method
4. Comprehend physical methods for the synthesis of nanoparticles

REVIEW QUESTIONS

1. Define nanotechnology and explain its significance in modern science and industry.
2. Describe the unique properties of materials at the nanoscale and explain how these properties differ from their bulk counterparts.
3. List and briefly describe the different types of nanomaterials, providing examples for each category.
4. Identify and discuss at least three major applications of nanotechnology, describing how nanomaterials are utilized in each field.
5. What are the primary reliability issues associated with MEMS and NEMS? Discuss the factors that contribute to these issues and potential strategies to mitigate them.

Answers to Multiple Questions

1. (a); 2. (c); 3. (b); 4. (a); 5. (a); 6. (b); 7. (b)

REFERENCES

1. Achanta, S., & Celis, J. P. (2015). Nanotribology of MEMS/NEMS. *Fundamentals of Friction and Wear on the Nanoscale*, *6*(1), 631–656.
2. Ali Mansoori, G., Bastami, T. R., Ahmadpour, A., & Eshaghi, Z. (2008). Environmental application of nanotechnology. *Annual Review of Nano Research*, *2*(1), 439–493.
3. Amen, R., Mukhtar, A., Saqib, S., Ullah, S., Al-Sehemi, A. G., Mehdi, S. E. H., & Bustam, M. A. (2021). History and development of nanomaterials. In *Nanomaterials: Synthesis, Characterization, Hazards and Safety* (Vol. *2*, pp. 1–14). Elsevier.
4. Arabi, S. (2023). Review on types of nanostructured materials. *International Journal of New Chemistry*, *11*(1), 48–57.
5. Arvidsson, R., Kushnir, D., Sandén, B. A., & Molander, S. (2013). How to make policy-relevant life cycle assessments of future products? Lessons learned from nanomaterials. In *6th International Conference on Life Cycle Management*, Gothenburg, *6*(3), 25–28.
6. Asiyanbola, B., & Soboyejo, W. (2008). For the surgeon: An introduction to nanotechnology. *Journal of Surgical Education*, *65*(2), 155–161.
7. Babu, V. J., Vempati, S., Uyar, T., & Ramakrishna, S. (2015). Review of one-dimensional and two-dimensional nanostructured materials for

INTRODUCTORY EXAMPLE

Nanoco Technologies is a prominent firm that focuses on the production and stability of semiconductor nanoparticles, which are commonly referred to as quantum dots. These nanoparticles possess distinctive optical and electronic characteristics that render them well-suited for use in display technologies, bio-imaging and solar cells.

Nanoco Technologies employs a chemical reduction technique to produce quantum dots that are free of cadmium. This procedure entails the decrease of metal precursors in combination with the presence of stabilizing chemicals to generate very radiant nanoparticles. To illustrate, the process of producing indium phosphide (InP) quantum dots involves reducing indium and phosphorus precursors under regulated conditions, leading to the formation of nanoparticles with exact dimensions and chemical makeup.

Nanoco utilizes organic ligands as surface capping agents to stabilize the quantum dots and inhibit their agglomeration. The ligands bind to the surface of the nanoparticles, creating steric hindrance that prevents the particles from coming into close contact. Ensuring the stability of quantum dots is essential for preserving their distinctive characteristics and guaranteeing their effectiveness in many applications.

Nanoco Technologies demonstrates the effective production and utilization of nanoparticles in cutting-edge applications through improved synthesis and stabilization processes. This highlights the significance of accurate control over nanoparticle properties in industrial and commercial settings.

UNIT INTRODUCTION

Metal atoms possess a significant level of chemical reactivity, which is maintained in their trimers, dimers, clusters, and nanoparticles that consist of a large number of atoms. Investigating active particles is feasible by employing different stabilizers (Lu et al., 2007). Consequently, it is essential to address the challenges related to synthesizing these nanoparticles and ensuring their stability simultaneously. Currently, there exist numerous techniques for producing particles of different sizes (Virkutyte & Varma, 2011).

Generally, the techniques used to create nanoparticles can be categorized into two main types. The initial category encompasses techniques that facilitate the manufacture and examination of nanoparticles, although provide limited assistance in the advancement of novel materials. The phenomena encompassed are condensation at extremely low temperatures, specific variations of chemical, radiation and photochemical reduction, as well as evaporation caused by the use of lasers (Gopalan, 2010). The second category consists of methods that facilitate the creation of nanomaterials and nanocomposites utilizing nanoparticles. These methods include gas phase condensation, mechanochemical dispersion, and plasmochemical synthesis, along with other comparable processes (Vollmer & Janiak, 2011). The earlier distinction illustrates another characteristic of the approaches being discussed, which may be described as follows: the particles can be formed either by assembling individual atoms (a "bottom-up" approach) or through different dispersion and aggregation techniques (a "top-down" approach). The "bottom-up" strategy primarily focuses on chemical techniques for producing nanoparticles, while the "top-down" approach is characteristic of physical methods. Figure 2.1 depicts both methods (Esumi, 2003). The process of synthesizing nanoparticles from individual atoms establishes the lower boundary of nanochemistry. The upper limit of this relates to atomic clusters, where the properties no longer experience significant changes as the number of atoms increases. This makes them similar to the qualities of dense metals (Sathiyanarayanan et al., 2017). Each element in the Periodic Table has a distinct number of atoms that determine its upper boundary. The architectures of nanoparticles of equal size can vary depending on the method used to obtain them, which is also highly significant. Typically, when compact materials are dispersed into nanosized particles, the resulting nanoparticles maintain their original structure (Verma et al., 2019). The electronic structure of particles created by the aggregation of atoms can be influenced by variations in the locations of the atoms. For instance, a particle with a size ranging from 2 to 4 nanometers can exhibit a reduction in the structural parameter (Laurent et al., 2008).

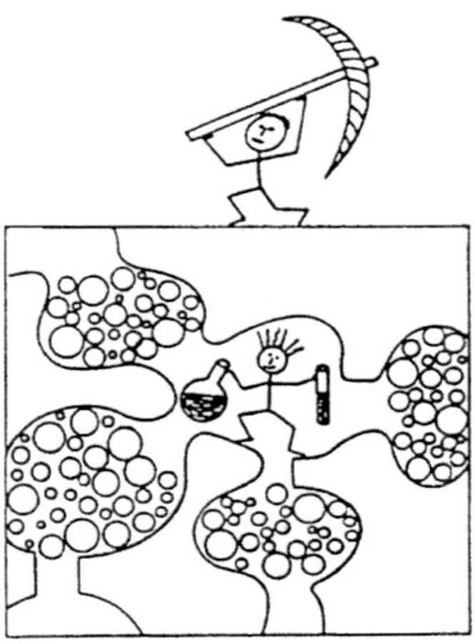

Figure 2.1. *Illustration of two approaches to the synthesis of nanoparticles.*

Source: G.B. Sergeev, Creative Commons License.

2.1. CHEMICAL REDUCTION

Learning Objectives

- Understand the principle of chemical reduction method
- Identify common reducing and stabilizing agents used in chemical reduction process

Currently, numerous experts are primarily focused on improving the techniques used in the manufacturing and maintenance of metal nanoparticles. Furthermore, particles that possess a consistent size are of great significance. The process of chemical reduction is mostly employed in the liquid phase, encompassing both aqueous and nonaqueous environments. Typically, metal compounds are distinguished by their salts. The extensive implementation of this method can be attributed to its ease and accessibility (Janiak, 2013).

To provide a visual representation, the procedure of generating gold particles is analyzed. A total of three solutions are prepared: (a) a solution of chloroauric acid in water, (b) a solution of sodium carbonate in water, and (c) a solution of hypophosphite in diethyl ether. Afterward, the mixture is exposed to a one-hour heating at a temperature of 70°C. As a result, gold particles ranging in size from 2 to 5 nm are obtained. A fundamental constraint of this method is the substantial existence of contaminants in a system of colloids consisting of gold nanoparticles. Nevertheless, this problem can be alleviated by utilizing hydrogen as the reducing agent. The behavior of a metal particle in a solution is mostly determined by the potential difference (ΔE), which represents the disparity between the electrode potential (E) and the redox potential (E_{redox}). The redox capacity of the particle at equilibrium is designated as E, whereas the potential of the equivalent solution is indicated as E_{redox}. Particles experience expansion when there is an increase in energy ($\Delta E > 0$) and disintegrate when there is a decrease in energy ($\Delta E < 0$). An unstable e□uilibrium occurs when there is no change in energy (ΔE). The complexity emerges due to the dependence of the redox potential of a metal particle on the number of atoms it contains. In such settings, chemical reduction is characterized by both thermodynamic and kinetic instability (Hu et al., 2015).

The technique where a reducer serves as both a stabilizer and a reducer has become popular recently. This category includes various nitrogen and sulfur-containing compounds, such as surfactants, thiols, nitrate salts, and polymers with functional groups. Alkali metal tetrahydroborates (MBH_4), which are suitable for use in acidic, neutral, and alkaline aqueous solutions, are the most widely used reagents for reducing metal ions. The alkali-metal tetraborates can reduce transition and heavy metal cations due to the relatively high redox potential of MBH_4 (1.24V in an alkaline environment) compared to the standard potentials of different metal ions, which fall within the range of 0.5 to 1.0V. It has been shown that complexes containing bridge bonds M—H—B are produced when metal ions are reduced. This enables the following transfer of a hydrogen atom, leading to the breaking of the bridge connection. Subsequently, a redox reaction occurs, leading to the cleavage of a B-H bond and the formation of BH_3 (Szczyglewska et al., 2023).

Various chemical substances and their derivatives, such as hydrazine hypophosphite, are frequently employed as reducing agents in the liquid media synthesis of metal nanoparticles. This section is focused on examining the precise aspects related to the speed and process of generating metal nanoparticles through redox reactions that take place in a liquid environment. A study was carried out by comparing the topochemical processes of heat solid degradation and the processes of crystallization, together with gas-solid reactions. Nevertheless, it is imperative to exercise prudence while examining the resemblances and outcomes derived from a formal depiction of the Kinetics of chemical reduction (Chaubey et al., 2007). For example, redox reactions are used to create, develop, and stabilize metal nanoparticles; further study is needed to understand the mechanisms and kinetics of these intricate and multifaceted processes. Within the reduced-metal-ion—reducer system, an electron is transferred from the reducer to the metal ion, causing chemical reactions. The electron transfer process is facilitated by the creation of an intermediate complex, resulting in a reduction in the required energy for the transfer (Peng et al., 2006).

With the use of sodium borohydride and tetraammonium disulfide, silver nitrate was reduced to create spherical silver nanoparticles with a diameter ranging from 3.3 to 4.8 nanometers. The stabilizer used was Dibromidebis[(trimethyl ammoniumdecanoylamino)ethyl]disulfide. The collected particles demonstrated significant light absorption at a wavelength of 400 nm, showing the existence of a silver plasmon peak and verifying the metallic composition of the particles. Particles tend to clump together in the presence of sulfuric and hydrochloric acids, according to research done on the effect of the medium on particle stability. The pH of the medium also had an impact on the durability of silver particles. The particles exhibited stability for one week in aqueous solutions with a pH range of 5–9. Silver particles accumulate and deposit more quickly when the pH is raised or lowered. The second factor had a rather minor effect on the stability of gold particles (De Souza et al., 2019).

Research indicates that the partial oxidation of the products obtained from reducing Ag^+ cations using borohydride in aqueous polyacrylate solutions can generate tiny silver clusters with a positive charge. These clusters are stabilized as polyacrylate complexes and are commonly referred to as "blue silver." The matrix was created with an amphiphilic polymer known as poly(octadecylsiloxane) to generate particles with exact sizes that range from 1 to 2 nanometers (Olad et al., 2017).

The reduction of various platinum salts was achieved by employing sodium borohydride and hydrazine as nanostructured media, together with polyelectrolyte gels including surfactants with opposite charges. The experiment demonstrated that sodium borohydride predominantly produces small platinum particles with radii of approximately 2–3 nm, whereas hydrazine forms particles measuring approximately 40 nm. The objective of the study was to investigate cobalt nanoparticle manufacturing mechanism, electron spectrum, and responses in water-based environments. Cobalt ions in aqueous solutions comprising $Co(ClO_4)_2$ and HCOONa were reduced by radiation-chemical methods to produce spherical cobalt particles with a diameter of 2–4 nanometers. A stabilizer was used, specifically sodium polyacrylate having a molecular mass of 2,100u. Radiolysis produces CO_2-radical ions, hydrogen atoms, hydroxyl radicals, and solvated electrons (e^-_{aq}) (Spadaro et al., 2010):

$$H_2O \rightarrow e^-_{aq}, H, OH,$$

$$OH(H) + HCOO^- \rightarrow H_2O(H_2) + CO_2^{-\bullet}$$

Radical ions and hydrated electrons Cobalt nanoparticles are produced via the reduction of CO_2+ ions, converting carbon dioxide (CO_2). The nanoparticles exhibit maximum absorbance at a wavelength of 200 nm. The autocatalytic nature of these activities was proven through the utilization of pulsed radiolysis. By subjecting aqueous solutions of $Ni(ClO_4)_2$ to radiation in the presence of isopropanol, polyacrylate, polyethylene, and polyvinyl sulfate, the Ni^{2+} ions were reduced, leading to the creation of metal sols (Aguilar et al., 2019). These sols are composed of spherical particles with diameters ranging from 2 to 4 nanometers. When nickel nanoparticles are coupled with silver ions, they form stable nickel-silver nanosystems. These nanoparticles react easily with both O_2 and H_2O_2. Copper particles with a spherical shape and a size ranging from 20 to 100 nm were generated by subjecting aqueous solutions containing $KCu(CN)_2$ to -radiolysis. To eliminate hydroxyl radicals, scavengers such as methanol or 2-propanol were used in this procedure (Manikam et al., 2011).

The study examined the production of silver particles by subjecting silver nitrate solutions to -radiolysis in water, ethanol, and a 0.01M $C1_2H_{25}OSO_3Na$ solution. In these solutions, the fractal dimensions of the particle clusters are 1.81, 1.73, and 1.70, respectively. The generation of stable ruthenium, rhodium, and platinum nanoparticles in an organic medium (average size: 1–2 nm) was reported in the study. The metal hydroxides were heated in ethylene glycol colloidal solutions to accomplish this (Pastoriza-Santos & Liz-Marzán, 1999).

Spherical amorphous selenium particles were synthesized in ethylene glycol by reducing selenious acid with hydrazine. Temperature adjustments between –10 and +60°C were used to create the customizable surfaces of the particles. The temperature range mentioned corresponds to the glass transition of selenium. This transition allows for the regulation of the pace at which iron oxide is incorporated into selenium particles (Guzmán et al., 2009).

Through the application of electrochemical methods, a silver anode (a plate made of silver) was dissolved in a solution of tetrabutylammonium bromide in acetonitrile, which lacks a proton-donating ability. This process resulted in the creation of silver nanoparticles with diameters that varied between 2 and 7 nanometers. This approach has demonstrated its dependence on variables such as the current density and the type of cathode. Therefore, when subjected to intense electric currents in non-equilibrium conditions, particles with unconventional forms can be generated. Increasing the current density from –1.35 to –6.90 milliamperes per square centimeter resulted in a drop in particle diameter from 6±0.7 to 1.7±0.4 nanometers. The addition of tetrabutylammonium bromide caused the silver ions to be reduced, resulting in the creation of metal nanoparticles. Afterward, these nanoparticles were utilized on cathodes composed of either platinum or aluminum. Figure 2.2 depicts this methodology. Platinum cathodes were coated with spherical silver nanoparticles. By depositing material on an aluminum cathode, films were produced. An examination of the optical spectra of nanoparticles during their creation has revealed the existence of an autocatalytic phase in this process. Furthermore, research has demonstrated that the plasmon band undergoes a downward shift in frequency as the size of the particle decreases. Additionally, the breadth of a peak associated with a particle's surface plasmon is directly proportional to the reciprocal of the particle's radius (denoted as R) (Dang et al., 2011).

When it comes to the production of nanoparticles, the utilization of organic solvents takes precedence. The functions that they conduct are stabilizing. When it comes to the process of synthesizing nanoparticles, solvents and surfactants are extremely important components. The polar groups can cling to the surface of developing nanocrystals, which allows them to form complexes with compounds that are present in solutions. This allows them to control the chemical reactivity of the substances and their migration towards the surface of the particle. The temperature, the surface energy of a nanocrystal, the concentration of free particles in solution, the particle sizes, and the surface-to-volume ratio are some of the critical parameters that impact the outcome of the processes. By regulating the aforesaid relationships, it becomes possible to control the generation of particles of different sizes. Recently, comparable techniques were employed to generate nanoparticles of many kinds, such as oxides like Fe_2O_3 and $MnFe_2O_4$, metal alloys like $CoPt_3$, and semiconductors like CdS, CdTe, InAs, InP, and Ge. The nanoparticles produced using these approaches exhibit rather limited variations in size (Lee et al., 2008).

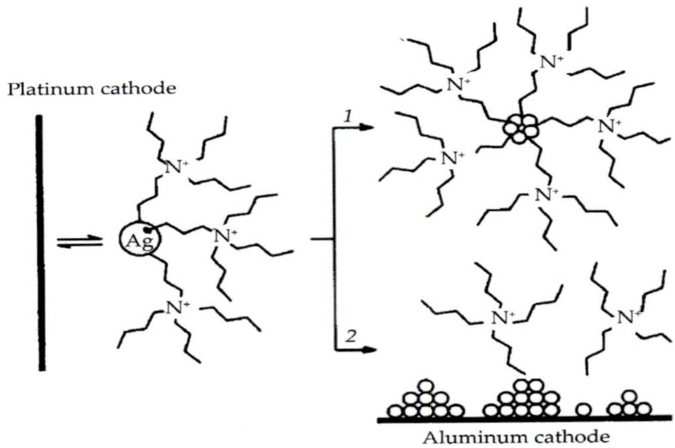

Figure 2.2. *Illustration of competition of two processes (1) formation of silver particles, (2) deposition of particles and film formation.*

Source: L. Blanko, Creative Commons License.

The preparation of various nanoparticles is accompanied by stabilization methods. To avoid aggregation, it is imperative to prohibit any interaction between active particles (Weare et al., 2000).

Did You Know?

The utilization of chemical reduction in the manufacture of nanoparticles can be traced back to prehistoric times. An illustration of this phenomenon can be observed in the Lycurgus Cup, an ancient Roman glass vessel dating back to the 4th century AD. The cup exhibits a remarkable ability to alter its hue depending on the angle of incident light, owing to the presence of gold and silver nanoparticles intricately incorporated throughout the glass structure.

2.2. PHOTOCHEMICAL AND RADIATION-CHEMICAL REDUCTION

Learning Objectives

- Comprehend the photochemical and radiation chemical reduction methods
- Understand the roles of light and radiation in nanoparticle formation

The synthesis of metal nanoparticles through the use of high energy affects the chemical system and leads to the production of highly reactive reducing agents such as electrons, radicals, and excited species. Photochemical reduction, also known as photolysis, and radiation-chemical reduction, also known as radiolysis, vary in terms of energy. Photosynthesis occurs at energy levels below approximately 60 electron volts (eV), while radiolysis operates at energy levels ranging from 103 to 104 eV. High-energy radiation processes are characterized by the following main features: non-equilibrium particle energy distributions; convergence of physical and chemical characteristics; time as a critical feature for chemical reactions involving reactive species and the presence of multi-channel and non-steady-state processes in the reacting systems (Singh et al., 2020).

There are several benefits associated with the photochemical and radiation-chemical reduction technologies in comparison to the chemical reduction approach. The usage of chemical reducers results in the production of nanoparticles that are very pure. This is because there is a minimal amount of contaminants produced throughout the process. Furthermore, the synthesis of nanoparticles can be accomplished in solid-state circumstances and at lower temperatures thanks to the synergy that exists between photochemical and radiation-chemical mechanisms. Photochemical reduction is now widely considered to be the most effective method for producing particles of noble metals in a solution. They were obtained from solutions that contained salts that were dissolved in water, alcohol, and organic solvents. These solutions were used to obtain these particles. The following is a list of the light-induced active species (Farooq et al., 2024):

$$H_2O \rightarrow e^-_{aq} + H + OH$$

When alcohols react, a hydroxyl radical and a hydrogen atom combine to form an alcohol radical:

$$H(OH) + (CH_3)_2CHOH \rightarrow H_2O(H_2) + (CH_3)_2COH$$

A solvated electron undergoes an interaction with a silver atom, causing a reduction in the silver atom's state to that of a metal:

$$Ag^{+} + e^{-}_{aq'} \rightarrow Ag^{0}$$

The UV absorption spectra exhibit peaks at 277 and 430 nm during the process of photoreduction, which indicates the presence of Ag^{4+} clusters and silver nanoparticles with sizes ranging from 2 to 3 nm, respectively. These peaks are observed during the early phase of photolysis. The maximum of the absorption band may shift toward either longer or shorter wavelengths as the amount of time that photolysis takes place increases. The presence of aggregation processes is indicated by the longwave shift, whilst the shortwave shift indicates a drop in the average size of silver particles. Both shifts demonstrate that silver particles are becoming smaller (Pomogailo et al., 2014).

The process of photoreduction results in the development of nanoparticles with specific dimensions as well as larger conglomerates, which are activated by light. To explore the effect of light, gold particles that were suspended in acetone, ethanol, and isopropanol were used as a specific example. The expansion and disappearance of the gold surface plasmon band at a wavelength of 523 nm was brought about by the utilization of a mercury lamp as a device for producing light. Consequently, the intensity of the band at 270 nm increased, and a distinct band emerged at 840 nm as a conse□uence of this. Based on the findings of the scientists, the alteration in the plasmon band was attributed to the dipole-dipole interaction that occurred between particles that were grouped. Immediately after being exposed to light for 20 hours, the gold particles proceeded through the process of full precipitation.

It has been noticed that the characteristics of the solvent and the wavelength of the light both play a role in determining the rate at which the particles aggregate. Ultraviolet radiation has a more significant impact than visible light. It has been determined that there is a connection between the wavelength of light and the intensification of van der Waals forces as well as the modifications in the interaction of Coulombs of surface charges that are brought about by light (Sakamoto & Majima, 2010).

The radiation-chemical reduction method for the manufacture of nanoparticles is gaining popularity as a result of its accessibility and reproducibility. In the process of forming metal nanoparticles in the liquid phase, the spatial organization of primary intermediate components is an extremely important factor. Radiolysis, in contrast to photolysis, results in the generation of intermediate particles that are scattered more uniformly. This makes it possible to acquire particles with narrower size distributions (Remita & Lampre, 2024).

Through the use of pulsed radiolysis, it was possible to create metal particles that possessed unique oxidation states. Following is the sequence of events that takes place during the reaction that involves the hydrated electron e_{aq}^{-}, which possesses a significant reduction potential (Rath, 2021):

$$M^{n+} + e_{aq}^{-} \rightarrow M^{(n-1)+}$$

As a result of the existence of a single electron in the outer orbital of an atom or metal ion, the high chemical reactivity of these entities is determined. The amount of time spent in water typically ranges from a few microseconds to a few milliseconds. The reduction potentials of these metal particles are what determine the optical qualities that they possess. For example, in the isoelectric series of metals that have electronic structures that are comparable to one another, the ionization potential rises as the charge of the metal ion rises. In addition, the wavelength at which the maximum amount of light is absorbed moves toward the shorter end of the range of wavelengths. Toward the beginning of the period, silver and gold atoms exhibit reductive properties, whereas toward the end of the period, trivalent tin and lead atoms, which have the same number of electrons, possess considerable oxidizing capabilities. A possible quantity of isocharged ions within the same subgroup is proportional to a rise in the period number, which corresponds to the increase (Sakamoto et al., 2009).

At the beginning of the radiation-chemical reduction process, atoms and extremely small metal clusters are generated. Subsequently, these atoms and clusters go through a change that results in the development of nanoparticles. They were stabilized by the utilization of additives that possessed features that were analogous to those employed in the process of chemical reduction. Through the usage of pulsed radiolysis and spectrophotometry, it was possible to investigate the first phases of the production of metal particles, with a particular emphasis on charged clusters such as Ag^{2+} and Ag^{4+}. The formation of metal nanoparticles took place as a result of the continuous contact that took place between clusters; however, the precise process that took place that led to this formation is still unknown.

Through the utilization of steady-state and pulsed radiolysis techniques, it was possible to facilitate the manufacture of a wide variety of nanoparticles that were composed of different metals. To manufacture bimetallic and trimetallic metal nanoparticles with a core-shell structure, new technologies that make use of radiolysis have been devised.

The goal of creating materials with various properties is particularly exciting, and nanoparticles that consist of several different metals are particularly fascinating. The reason for this is that it is possible to form intermetallic compounds and alloys at the nanoscale, which are not accessible in their solid metal state (Bárta et al., 2019).

The salt solutions were subjected to radiation-induced reduction, which resulted in the formation of nanoparticles that were composed of two or three different metals. Au-Hg particles were produced by a process that consisted of two stages. Initially, the radiation-chemical reduction technique was utilized in order to produce gold nanoparticles with a size of 46 nanometers. Following that, $Hg(ClO_4)_2$ and isopropanol were added to the gold sol, which resulted in the precipitation of mercury ions onto the gold particles. After that, the mercury ions proceeded through a process of reduction as a result of the action of the free radicals that were created during the radiolysis process. The reduction of Na_2PdCl_4 salt resulted in the production of palladium nanoparticles with a diameter of 4 nanometers and a narrow size distribution (Shirokova et al., 2016). Sodium citrate served as a stabilizer during this preparation process. The reduction of the gold ions occurred after the addition of $K_2Au(CN)_2$ to a methanol solution that included palladium particles, which was then followed by exposure to radiation. During the entirety of the process, there was no trace of individual gold particles being produced. Instead, all of the gold was placed uniformly onto the palladium particles, which resulted in the formation of a coating on the surface of the palladium. In addition to this, it is possible to apply a silver coating on the Pd—Au particles. The created particles were made up of palladium cores that were encased in two different shells, which were gold and silver respectively. When it comes to the study of ultrafast electrical events, these multiplayer clusters are absolutely exciting (Dispenza et al., 2015).

Through the utilization of silicates that have been changed with organic compounds that serve as matrix and stabilizers, it is possible to produce solvents and gels, in addition to bimetallic nanoparticles, in a single step. According to the findings of tests that were carried out with scanning tunneling microscopy (STM), researchers have shown that Pt—Pd particles are composed of palladium cores that are surrounded by platinum shells. Utilizing thin silicate sheets that were embedded with bimetallic nanoparticles allowed for the electrocatalytic oxidation of ascorbic acid to be carried out. The second example demonstrates that the combination of two metals can significantly and fre□uently in an unpredictable manner alter the properties of nanoparticles. Through the use of an additional metal on top of a conventional metallic catalyst, it is possible to develop one-of-a-kind surfaces that display creative properties (Egorova et al., 2016).

Gold and palladium nanoparticles with a bimetallic structure were created by the use of the sonochemical technique. Combining aqueous solutions of $NaAuCl_4.2H_2O$ and $PdC1_2{}^-2NaCl^-3H_2O$ with sodium dodecyl sulfate, which acted as a reducing and stabilizing agent, resulted in the formation of the particles. Within the core of the bimetallic particles, gold atoms were included, and palladium atoms were incorporated into the surrounding shell. To establish the diameters of the cores and shells for a variety of gold-to-palladium ratios, high-resolution electron microscopy provided the necessary information (Zharikov et al., 2020). Through the use of X-ray spectroscopy, it was possible to ascertain the respective amounts of gold and palladium within the particles. For both the cores and the shells, it is possible to make an approximation of their dimensions by making use of the information that has been provided regarding density, mass, and initial ratios. In Table 2.1, a comparison is made between the data obtained from experiments and the data that was computed. The table demonstrates that the sonochemical manufacturing of bimetallic particles enables the control of core and shell diameters by adjusting the quantities of Au^{3+} and Pd^{2+}. This is accomplished through the use of the sonochemical process (Pietrobon & Kitaev, 2008).

Table 2.1. Representation of Core (Gold) Diameter and Shell (Palladium) Thickness in Bimetallic Au-Pd Particles, nm

AU/Pd	**Experiment**		**Calculations**	
Ratio	Core diameter	Shell thickness	Core diameter	Shell thickness
1:1	6.0	1.0	6.4	0.8
1:4	5.0	1.5	4.8	1.6

Source: G. B. Sergeev, Creative Commons License.

Fe-TiH particles were created using mechanosynthesis by combining iron and titanium hydride. Nanoscale iron-tungsten compositions, with tungsten content ranging from 2 to 85 atomic percent, were synthesized by reducing a mechanical mixture of iron hydroxide ($FeOOH.nH_2O$) and tungstic acid (H_2WO_4) in a hydrogen atmosphere at a temperature of 740°C for one hour. The particles were examined utilizing X-ray diffraction (XRD) and Mössbauer spectroscopy techniques. The sol-gel synthesis is classified into different forms according to the chemical processes employed for the production of oxide and sulfide nanoparticles. The procedure for synthesizing metal nano oxides is as outlined below (Saha et al., 2010):

$$M(OR)_n + xH_2O \rightarrow M(OH)_x(OR)_{n-x} + xROH \quad \text{(hydrolysis)}$$

$$M(OH)_x(OR)_{n-x} \rightarrow MO(n/2)_x + (2x - n)/2\, H_2O + (n-x)\, ROH \quad \text{(condensation)}$$

M denotes a metallic element, whereas R signifies an alkyl group. The process was expedited by the modifications in the initial solution's pH. Linear chains were detected in acidic settings, whereas branching chains were seen in alkaline solutions. To ac□uire metal sulfide nanoparticles, it is possible to substitute the alkoxides of metals with the corresponding sulfides M(SR) and thereafter subject them to hydrosulfide treatment. Now, researchers are actively working on developing new methods to create nanocrystalline oxide materials by utilizing different chemicals at subcritical and supercritical conditions. The most frequently employed constituents are carbon dioxide and, more recently, water. By dissolving metal oxide particles in electrolytes, nanoparticles were formed, reducing their size from the micron to the nanoscale (Marin et al., 2008).

Did You Know?

Gold nanoparticles exhibit different colors, such as red or purple, based on their dimensions and morphology. The observed shift in hue is a result of surface plasmon resonance, which occurs when the electrons that conduct on the surface of the nanoparticle vibrate in sync with light waves.

Metal, semiconductor, and their complexes with biomolecule nanoparticles were produced employing a process that involved quickly expanding supercritical solutions into a liquid solvent. This method was utilized to generate particles with an average diameter of less than roughly 50 nm by utilizing polymers that are soluble in CO_2. The technique of quickly expanding supercritical solutions in li□uids was utilized to generate finely divided nonsteroidal anti-inflammatory drugs, namely Ibuprofen and Naproxen. The solution was rapidly diluted in water using either poly(N-vinyl-2-pyrrolidone) with a molecular weight of approximately 40,000u (Zezin et al., 2020).

2.3. CRYOCHEMICAL SYNTHESIS

Learning Objectives

- Understand the principle of cryochemical synthesis
- Comprehend the challenges associated with producing nanoparticles at low temperatures

Small clusters and metal atoms with high activity aggregate into bigger particles when stabilizers are not present. The aggregation process occurs without the need for activation energy. The approach of matrix isolation was used to achieve the stabilization of active atoms from nearly all elements in the Periodic Table at low (77 K) and super low (4–10 K) temperatures. This technology relies on the utilization of inert gasses at extremely low temperatures. The gasses typically used as matrices include argon and xenon. Metal atoms in the form of vapors are combined with a significantly larger amount of inert gas on a surface that is cooled to a temperature range of 4–12

K. The significant reduction in concentration due to the presence of an inactive gas and the extremely low temperatures effectively prevent the movement of metal atoms through diffusion; as a result, they are kept stable in the condensed state. Spectral and radiospectral methods are used to study the physicochemical features of these atoms (Shlyakhtin & Tretyakov, 1999).

For a chemical reaction to take place at low temperatures, it is necessary for the active particles that are stabilized in the condensate to be capable of movement. Matrix separation and chemical reactions are fundamentally incompatible processes. Active particles become stabilized when their reactions cease, and conversely, the existence of a chemical reaction indicates the absence of stabilization (Sergeev & Shabatina, 2008). When examining samples acquired using the matrix isolation technique, subjecting them to heat revealed numerous novel and unconventional chemical interactions between atoms and specific chemical compounds, particularly when placed into low-temperature condensates.

These types of transformations can be represented by the following scheme (Trusova & Trutnev, 2018):

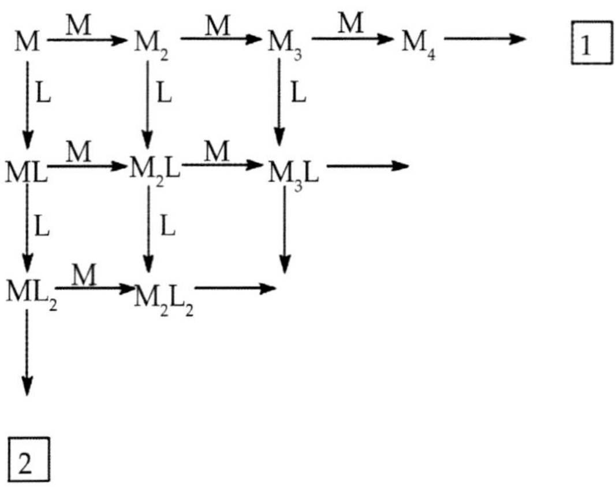

M denotes a metal, whereas L represents a ligand, which is a chemical molecule. This system demonstrates a sequence of consecutive and concurrent competitive responses. Direction 1 entails the process of combining metal atoms to create dimers, trimers, and nanoparticles. Direction 2 deals with the atom-ligand interaction that produces complexes, or organometallic compounds. The activities taking place in low-temperature condensates, as described by the aforementioned framework, are marked by non-equilibrium circumstances and are affected by multiple parameters. The technique of producing metal nanoparticles through co-condensation on low-temperature surfaces allows for the inclusion of various additives, which can modify the physical and chemical properties of the system (Morozov et al., 2017).

Recognizing certain features is crucial when constructing materials for cryochemical reactions involving metal nanoparticles, atoms, and clusters. Metal atoms can be generated by several heating techniques. Alkali, alkaline earth, and certain other metals exhibit a strong tendency to evaporate. Their vapors can be obtained by directly heating them. Heating is usually provided by using a low-voltage (5 V) transformer that offers high current densities (up to 300A). There is cooling occurring in the inlet tubes. The metal sample that is going to be transformed into vapor can be shaped into many forms, such as a wire, spiral, or ribbon. Usually, metals with high conductivity like copper (Cu), silver (Ag), and gold (Au) are transformed into vapor in Knudsen cells using either direct or indirect heating techniques. By employing the variables of temperature (T), pressure (P), and size of the exit aperture, the rate of evaporation can be computed (Sergeev et al., 1999):

$$N = P/(2\pi MRT)^{1/2}$$

The universal gas constant is denoted by R, the molecular mass is represented by M, and the rate of substance evaporation per second per square centimeter of the outlet hole area is indicated by N.

Numerical methods can be employed to solve the equations governing temperature changes, enabling the computation of the temperature distribution within the Knudsen-cell chamber. Indirect heating provides a higher level of temperature uniformity throughout the cell. Optical pyrometers are employed to ascertain the temperature at which a metal undergoes evaporation. The condensate temperature is monitored using thermocouples (Trakhtenberg & Gerasimov, 2005).

Needle valves are employed to control the passage of gasses or vapors by manipulating the pressure drop in a container of a specified volume. The choice of valve materials is based on the properties of the chemicals being processed and their susceptibility to corrosion. Multiple rheometers and Bernoulli pressure gages are employed to quantify the reduction in pressure within a capillary as gas flows through it. The reduction in pressure is closely correlated with the velocity of the flow and is □uantified using pressure gages. The flow velocities used in the matrix isolation approach range from approximately 0.1 to 0.01 mmol/h. Over several hours, the temperature of a surface where a sample is condensed rises gradually by tiny degrees at these deposition rates. When analyzing the chemical processes in low-temperature condensates, it is essential to guarantee that no chemical reactions occur during the synthesis of the sample. By using a molecular beam mode, the gas-phase processes are eliminated. To avoid collisions and gas-phase reactions, the free route λ should exceed the dimensions of the cryoreactor. The □uantity λ exhibits an inverse relationship with the pressure, as expressed by the e□uation λ = K/p, where K represents a constant (Yelenich et al., 2015).

When utilizing a Knudsen cell, it is crucial to take into account the Knudsen number, which is computed as 2λ/R, with R denoting the characteristic dimension. Pressures in the range of 103 mmHg are sufficient when the wavelength (λ) is much greater than the radius (R) and the Knudsen number (Kn) is much more than 1. At a temperature of T=300K, the flow undergoes a transition from a continuous mode to a molecular one when the wavelength (λ) is significantly less than the characteristic length (R) and the Knudsen number (Kn) is significantly less than 1. The rate of particle evaporation from a Knudsen cell can be calculated by consulting tables that present the correlation between vapor pressure and temperature. A straightforward technique for determining the amount of metal that has vaporized entails measuring the weight of the metal sample before and after the experiment (Shabatina & Sergeev, 2010). When all evaporated substances are contained inside a sample's transmission and absorption bands, optical methods can be used. The quartz crystal microbalance method is used to measure the amount of metal that has been deposited. This microbalance depends on the linear relationship between the mass of the material being deposited and the frequency of vibrations of the quartz crystal. Standard quartz crystals have a frequency of 5,000 kHz, a diameter of 1 mm, and a thickness of 0.3 mm. An additional mass is introduced, and this shifts the resonance fre□uency. There are five times as many nanograms per hertz as there are crystals. The following equation establishes the link between the mass m and the resonance frequency f (Trakhtenberg et al., 2002):

$$\Delta f - f K \Delta m/S$$

A quartz crystal's density (2.65 g/cm^2) and surface area (S) define the fre□uency shift (Δf). The constant (K) is dependent on these two factors. The crystal is firmly attached to a support between two generator electrodes. Before use, the balance is calibrated, and the frequency meter logs resonance frequency fluctuations (Vernaya et al., 2017).

Several specialized cryoreactors were specifically constructed for cryochemical synthesis of atoms, clusters, and nanoparticles. Figure 2.3 illustrates a cryoreactor used for performing matrix isolation and spectroscopic studies on reactive metals at temperatures ranging from 12 to 70 K. Figure 2.4 illustrates a cryoreactor configuration used for condensation at temperatures equal to or over the boiling point of liquid nitrogen (77 K). The main element of this unit is a purified copper cube that has been pre-cooled to a temperature of 77 K. Once the compounds being studied are applied, the sample is rotated by 180° and infrared reflectance spectra are documented. The cryoreactor allows for the collection of spectra at different temperatures while ensuring precise control over the sample temperature. UV and optical transmission spectra can be ac□uired by employing salt and □uartz windows affixed within hollow copper holders (Generalov & Trutnev, 2007).

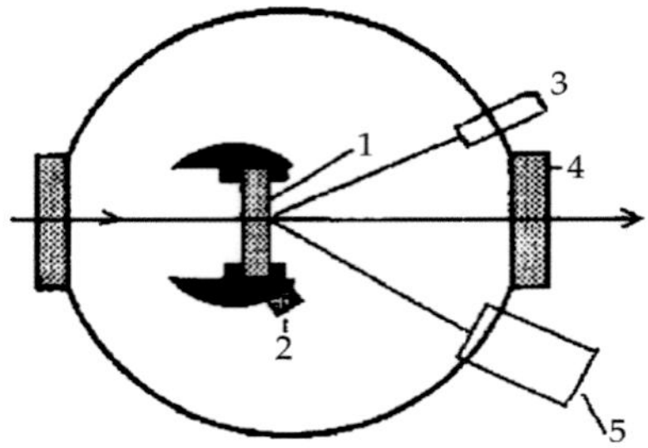

Figure 2.3. Schematic of window-level cross-section of a cryostat designed for temperature range of 1*2–70 K;(1) window with a formed sample, (2) quartz microbalance, (3) gas inlet, (4) external windows, and (5) evaporator of metal.*

Source: A. Galkin, Creative Commons License.

Specialized cryoreactors were designed for the special purpose of obtaining electron paramagnetic resonance (EPR) spectra. Figure 2.5 depicts the schematic diagram of a cryoreactor utilized in the thin film material production process. As an example, a setup developed at Oxford University is presented (Figure 2.6). One benefit of this technology is the capacity to produce grams of pharmaceuticals in just a few hours of operation. This arrangement was widely employed for the synthesis of compounds incorporating rare-earth elements (REE). Japanese scientists utilized a simplified approach. The ligand undergoes sublimation and is then coated onto the inner surface of a cryogenic vessel by utilizing liquid nitrogen for cooling. After 60 minutes of conditioning, the condensate is slowly heated and then analyzed. Usually, this approach allows for the acquisition of new chemical compounds (Balakhonov et al., 2018).

Figure 2.7 illustrates a reactor capable of facilitating the chemical reaction of a small amount of material, specifically a few milligrams. The condensation of metal vapors, ligands, and, if necessary, stabilizers takes place on the glass vessel walls, which are chilled by liquid nitrogen. Once the condensation process is finished, the sample is exposed to heat and gathered at the base of the container. This enables the collection of the sample for further analysis without interrupting the suction. As a result, a variety of sols or organodispersions of metals were successfully obtained. The cryostat was modified to obtain systems of nanoparticles consisting of two different metals (Saha et al., 2019). Similar, albeit partially modernized, reactors were used to condense metal vapors into extremely cold liquids at the bottom of a cryostat. One benefit of these reactors is the relatively simple nature of research. Various new organometallic compounds were synthesized in similar reactors through the simultaneous or sequential condensation of metal vapors with various ligands. However, the cryoreactors stated previously encounter the difficulty of precisely measuring the reagent ratio, which is a significant drawback. The Green cryoreactor, fre□uently utilized for cryosyntheses (Figures 2.8c and d), possesses a comparable drawback. Essentially, it operates as a rotary evaporator with a flask that rotates and is immersed in a chilled bath (Kurapova et al., 2012).

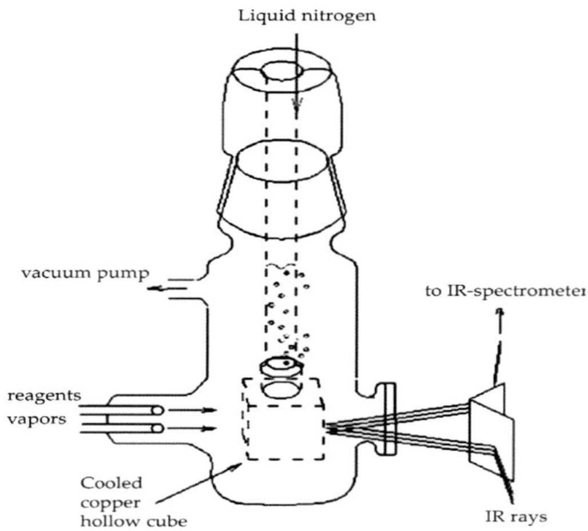

Figure 2.4. *Cryostat for low-temperature IR spectroscopy at 77 k.*

Source: G. B. Sergeev, Creative Commons License.

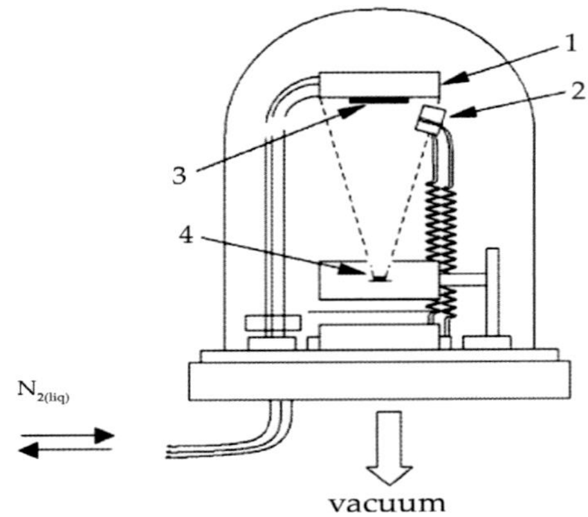

Figure 2.5. *Illustration of experimental setup scheme for low-temperature physical vapor deposition.*

Source: P.L. Arnold, Creative Commons License.

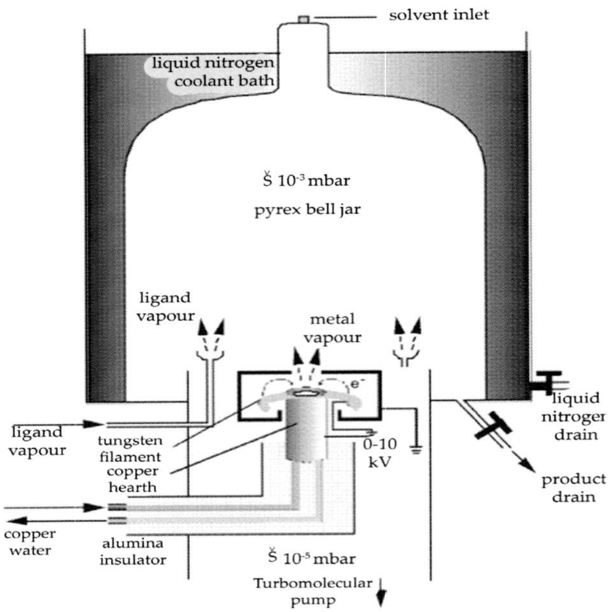

Figure 2.6. *Unit for preparative synthesis of organometallic compounds by cryo-condensation.*

Source: P.L. Arnold, Creative Commons License.

Nanophase Technologies Corp. (USA) suggested a fascinating cryochemical synthesis method for composition materials in the early 1990s (Gerasimov et al., 1996).

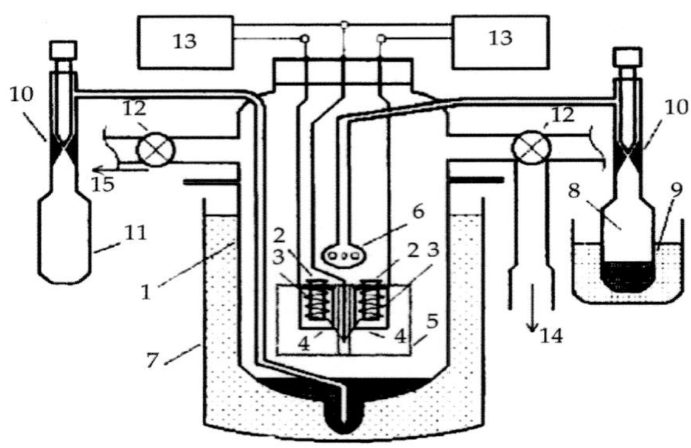

Figure 2.7. *Scheme of semipreparative reactor for cryochemical reactions.*

Source: K. Mochida, Creative Commons License.

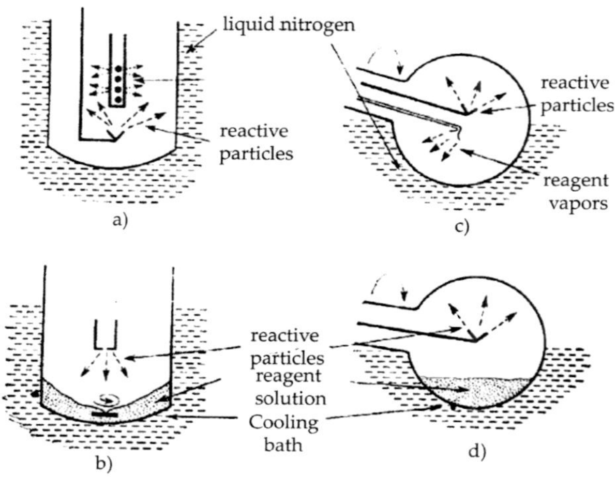

Figure 2.8. *Scheme of reactors for preparative cryonanochemistry.*

Source: Benfield, Creative Commons License.

Hexogen ($C_3H_6N_3(NO_2)_3$), ammonium nitrate, and its mixtures are currently produced as nanoparticles using the low-temperature co-condensation process. The analysis using atomic force microscopy revealed that the mixture is composed of ammonium nitrate particles measuring 50 nm and hexogen particles measuring 100 nm. The authors did not provide precise information regarding the modifications in the characteristics of high-energy systems due to variations in particle size. Scientists sometimes face a dilemma known as the "macro-micro" issue when using cryochemical and other methods to produce metal nanoparticles. The challenge arises from the existence of particles with unique sizes and distinct chemical processes encountered when handling samples, particularly those generated in reactors designed for spectral analysis. Nevertheless, attempts to reproduce the results by using larger amounts of chemicals often lead to the formation of particles with different sizes and uni□ue chemical processes. In contrast, there are situations where a scientific investigation carried out on a large scale cannot be duplicated on a very small size. Resolving this contradiction is one of the challenging conundrums in the domain of chemistry, particularly in the arena of nanochemistry (Rothgery et al., 1975).

Practice Problem

A researcher is employing laser ablation to fabricate titanium dioxide (TiO_2) nanoparticles. Explain the procedure and determine the essential variables that require regulation during the synthesis.

Solution to Practice Problem

Process:

1. Immerse a large titanium target in a liquid media, such as water or ethanol.
2. Direct a pulsed laser beam onto the titanium target to remove material by ablation.
3. The material that has been removed by ablation transforms into a plasma plume, which subsequently cools down and condenses. This condensation process leads to the formation of TiO_2 nanoparticles inside the liquid medium.
4. Retrieve and refine the nanoparticles through the process of centrifugation or filtration.

Key Parameters:

- Laser Wavelength: The laser wavelength must be appropriate for the substance being ablated. Titanium is typically employed with wavelengths in the ultraviolet (UV) to visible range, such as 355 nm or 532 nm.
- Pulse Duration and Repetition Rate: Controlling the energy given to the target and the size of the nanoparticles can be achieved by using shorter pulse durations (such as nanosecond or femtosecond) along with appropriate repetition rates.
- Laser Energy Density: The energy concentration (fluence) should be enough to remove the target material without producing undue heating or damage.
- Ambient Medium: The selection of a liquid medium has an impact on the rate of cooling and stabilization of nanoparticles.
- Target Material Properties: The degree of cleanliness, state of the surface, and thickness of the titanium.

2.4. PHYSICAL METHODS

Learning Objectives

- Comprehend the physical methods and their mechanisms for nanoparticle synthesis

Diverse physical methodologies can be utilized to produce metal nanoparticles (Dhand et al., 2015). A key method includes combining the vaporization of metal with a flow of inert gas and then causing the resulting mixture to condense in a chamber maintained at a precise temperature (Iravani et al., 2014). Several versions of this technique were carefully analyzed. Physical methods such as gas evaporation, shock waves, electro explosion, laser-induced electrodispersion, low-temperature plasma, molecular beams, cathode sputtering, and other mechanical dispersions are used in traditional methods of nanoparticle fabrication (Ansari et al., 2022).

Figure 2.9 depicts the initial arrangement used to create metal nanoparticles with a high level of porosity. The operation of this arrangement relies on a closed gas cycle, where particles of metals such as silver are deposited onto a filter and subse□uently dislodged by gas pulses (Fuskele & Sarviya, 2017). This enables the execution of an almost uninterrupted procedure and the acquisition of adequately sizable porous particles through the aggregation of nanoparticles. Figure 2.10 depicts the arrangement used to create metal-polymer compounds by plasma jet synthesis (Rane et al., 2018).

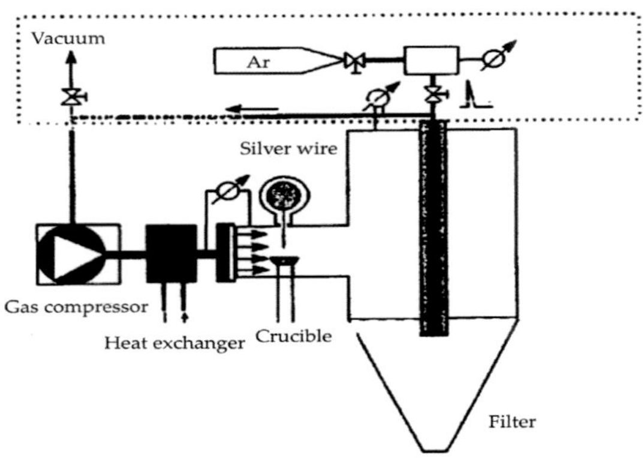

Figure 2.9. *Illustration of a setup for continuous production of highly porous metals as a result of aggregation of metal nanoparticles.*

Source: H. G. Busman, Creative Commons License.

The setup is divided into several zones for the plasma-induced production of nanoparticles and the subsequent coating process with monomers. Chlorides, carbonyls, and organometallic compounds are injected into the discharge zone along with helium as precursors (Thakkar et al., 2010). The particles acquire an electric charge that hinders their collision and the subsequent creation of clusters. This module facilitated the manufacture of metal oxides, nitrides, sulfides, and carbide particles, ranging in size from 5 to 20 nanometers, which were coated with polymers (Kraynov & Müller, 2011).

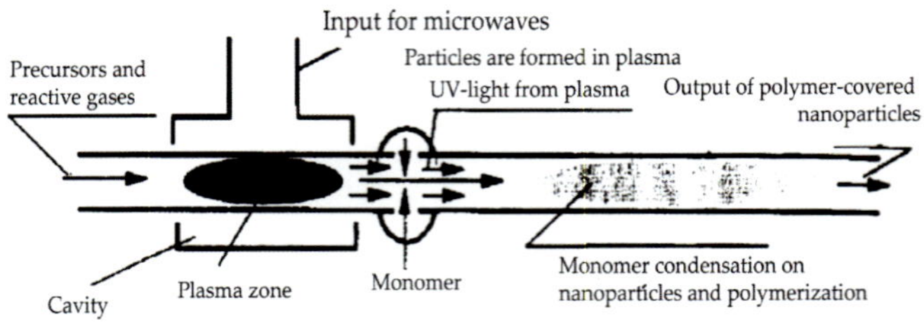

Figure 2.10. *Unit for continuous production of nanoparticles encapsulated in polymers.*

Source: S. Y. Kim, Creative Commons License.

New methods were developed that use laser evaporation to apply coatings on various particles, as well as a combination of combustion and chemical deposition (Nagasawa et al., 2002).

Condensation approaches include the physical methods for creating nanoparticles that were previously mentioned. Multiple versions of mechanical dispersion have also been extensively employed (Razi et al., 2022). The latter methods exhibit similarities to chemical methods in certain aspects. Several specialized publications that were carefully reviewed went into great detail about the mechanochemical dispersion technique used to create nanoparticles (Ealia & Saravanakumar, 2017). Presently, the primary objective in the production of nanoparticles from different elements is to advance technologies that facilitate the creation and preservation of particles that are 1 nm or smaller in size (Toshima & Yonezawa, 1998).

Did You Know?

Laser ablation can produce nanoparticles with high purity since it does not involve the use of chemical reagents that may introduce contaminants.

SUMMARY

- This chapter offers a thorough examination of the several techni□ues employed in the production and preservation of nanoparticles; encompassing both chemical and physical methodologies, emphasizing their fundamental concepts, procedures, and practical uses.
- The chapter explores the process of chemical reduction, which entails reducing metal ions in a solution to create nanoparticles. Typical examples of reducing agents are sodium borohydride, hydrazine, and citrate. Stabilizing compounds are commonly used during the process to prevent the clumping together of particles and to regulate their size.
- It also emphasizes the utilization of photochemical and radiation-chemical reduction techniques. Photochemical reduction employs light to catalyze the reduction of metal ions into nanoparticles. Radiation-chemical reduction utilizes high-energy radiation, such as gamma rays, to activate the reduction process.
- The Cryochemical Synthesis approach entails the production of nanoparticles at extremely low temperatures. Cryochemical techniques can generate nanoparticles that are exceptionally consistent in size and shape. The low-temperature environment facilitates the stabilization of reactive intermediates and inhibits undesired side reactions.
- Lastly, the chapter discusses physical techni□ues such as physical vapor deposition and laser ablation. These methods are commonly employed to generate nanoparticles from large quantities of materials using mechanical or thermal techniques. Physical techniques can attain a high level of purity and precise control over the content and structure of nanoparticles.

MULTIPLE CHOICE QUESTIONS

1. Which of the following is a common reducing agent used in the chemical reduction method for nanoparticle synthesis?

a. Sodium chloride

b. Sodium borohydride

c. Ethanol

d. Acetone

2. In photochemical reduction, what is primarily used to drive the reduction of metal ions to form nanoparticles?

a. Heat

b. Pressure

c. Light

d. Magnetic fields

3. Which method involves the use of high-energy radiation to initiate the reduction process in nanoparticle synthesis?

a. Chemical reduction

b. Cryochemical synthesis

c. Photochemical reduction

d. Radiation-chemical reduction

4. Cryochemical synthesis of nanoparticles is typically conducted at:

a. High temperatures

b. Room temperatures

c. Low temperatures

d. Variable temperatures

5. What is the primary advantage of cryochemical synthesis in nanoparticle production?

a. High temperatures

b. Uniform particle size and morphology

c. Fast reaction rates

d. Low cost

REVIEW QUESTIONS

1. Describe the chemical reduction method for nanoparticle synthesis. What are some common reducing agents used in this process?
2. Explain the process of photochemical reduction. How does light intensity influence the size and distribution of nanoparticles?
3. Discuss radiation-chemical reduction. What types of high-energy radiation are typically used, and what are the advantages of this method?
4. What is cryochemical synthesis? Highlight the benefits and challenges associated with synthesizing nanoparticles at low temperatures?
5. Explain how the choice of synthesis method can influence the size, shape, and uniformity of nanoparticles?

Answers to Multiple Questions

1. (b); 2. (c); 3. (d); 4. (c); 5. (b)

REFERENCES

1. Aguilar, M. S., Esparza, R., & Rosas, G. (2019). Synthesis of Cu nanoparticles by chemical reduction method. *Transactions of Nonferrous Metals Society of China*, *29*(7), 1510–1515.
2. Ansari, M. J., Kadhim, M. M., Hussein, B. A., Lafta, H. A., & Kianfar, E. (2022). Synthesis and stability of magnetic nanoparticles. *BioNanoScience*, *12*(2), 627–638.
3. Balakhonov, S. V., Teben'kov, P. V., Brylev, O. A., & Churagulov, B. R. (2018). Comparison of the physicochemical and electrochemical properties of vanadium oxide-based nanomaterials prepared by cryochemical synthesis and supercritical drying technique. *Inorganic Materials*, *54*, 60–65.
4. Bárta, J., Procházková, L., Vaněček, V., Kuzár, M., Nikl, M., & Čuba, V. (2019). Photochemical synthesis of nano-and micro-crystalline particles in aqueous solutions. *Applied Surface Science*, *479*, 506–511.
5. Chaubey, G. S., Barcena, C., Poudyal, N., Rong, C., Gao, J., Sun, S., & Liu, J. P. (2007). Synthesis and stabilization of FeCo nanoparticles. *Journal of the American Chemical Society*, *129*(23), 7214–7215.
6. Dang, T. M. D., Le, T. T. T., Fribourg-Blanc, E., & Dang, M. C. (2011). Synthesis and optical properties of copper nanoparticles prepared by a chemical reduction method. *Advances in Natural Sciences: Nanoscience and Nanotechnology*, *2*(1), 15009.
7. De Souza, C. D., Nogueira, B. R., & Rostelato, M. E. C. (2019). Review of the methodologies used in the synthesis of gold nanoparticles by chemical reduction. *Journal of Alloys and Compounds*, *798*, 714–740.
8. Dhand, C., Dwivedi, N., Loh, X. J., Ying, A. N. J., Verma, N. K., Beuerman, R. W., & Ramakrishna, S. (2015). Methods and strategies for the synthesis of diverse nanoparticles and their applications: A comprehensive overview. *RSC Advances*, *5*(127), 105003–105037.
9. Dispenza, C., Grimaldi, N., Sabatino, M. A., Soroka, I. L., & Jonsson, M. (2015). Radiation-engineered functional nanoparticles in aqueous systems. *Journal of Nanoscience and Nanotechnology*, *15*(5), 3445–3467.
10. Ealia, S. A. M., & Saravanakumar, M. P. (2017). A review on the classification, characterization, synthesis of nanoparticles and their application. In *IOP Conference Series: Materials Science and Engineering*. IOP Publishing. (Vol. *263*, pp. 32019–32025).

11. Egorova, E. M., Kubatiev, A. A., & Schvets, V. I. (2016). Methods of the nanoparticle preparation in solutions. *Biological Effects of Metal Nanoparticles*, *3*, 70–77.

12. Esumi, K. (2003). Dendrimers for nanoparticle synthesis and dispersion stabilization. *Colloid Chemistry*, *2*, 31–52.

13. Farooq, N., Taha, A., & Hashmi, A. A. (2024). Photoinduced synthesis of nanoparticles. *Green Synthesis of Nanomaterials: Biological and Environmental Applications*, *7*(2), 89–113.

14. Fuskele, V., & Sarviya, R. M. (2017). Recent developments in nanoparticles synthesis, preparation and stability of nanofluids. *Materials Today: Proceedings*, *4*(2), 4049–4060.

15. Generalov, M. B., & Trutnev, N. S. (2007). Cryochemical method of fabricating nanomaterials. *Theoretical Foundations of Chemical Engineering*, *41*(5), 628–633.

16. Gerasimov, G. N., Sochilin, V. A., Chvalun, S. N., Volkova, L. V., & Kardash, I. Y. (1996). Cryochemical synthesis and structure of metal-containing poly (p-xylylenes): System poly (chloro-p-xylylene) -Ag. *Macromolecular Chemistry and Physics*, *197*(4), 1387–1393.

17. Gopalan, P. R. (2010). Cyclodextrin-stabilized metal nanoparticles: Synthesis and characterization. *International Journal of Nanoscience*, *9*(05), 487–494.

18. Guzmán, M. G., Dille, J., & Godet, S. (2009). Synthesis of silver nanoparticles by chemical reduction method and their antibacterial activity. *International Journal of Chemical and Biomolecular Engineering*, *2*(3), 104–111.

19. Hu, H., Xin, J. H., Hu, H., Wang, X., Miao, D., & Liu, Y. (2015). Synthesis and stabilization of metal nanocatalysts for reduction reactions–A review. *Journal of Materials Chemistry A*, *3*(21), 11157–11182.

20. Iravani, S., Korbekandi, H., Mirmohammadi, S. V., & Zolfaghari, B. (2014). Synthesis of silver nanoparticles: Chemical, physical and biological methods. *Research in Pharmaceutical Sciences*, *9*(6), 385–406.

21. Janiak, C. (2013). Ionic liquids for the synthesis and stabilization of metal nanoparticles. *Zeitschrift für Naturforschung B*, *68*(10), 1059–1089.

22. Kraynov, A., & Müller, T. E. (2011). Concepts for the stabilization of metal nanoparticles in ionic liquids. *Applications of Ionic Liquids in Science and Technology*, *9*, 235–60.

23. Kurapova, O. Y., Konakov, V. G., Golubev, S. N., Ushakov, V. M., & Archakov, I. Y. (2012). Cryochemical methods for manufacturing nanosized

ceramics and ceramic precursor powders with low agglomeration degree: A review. *Reviews on Advanced Materials Science*, *32*, 112–132.

24. Laurent, S., Forge, D., Port, M., Roch, A., Robic, C., Vander Elst, L., & Muller, R. N. (2008). Magnetic iron oxide nanoparticles: Synthesis, stabilization, vectorization, physicochemical characterizations, and biological applications. *Chemical Reviews*, *108*(6), 2064–2110.
25. Lee, Y., Choi, J. R., Lee, K. J., Stott, N. E., & Kim, D. (2008). Large-scale synthesis of copper nanoparticles by chemically controlled reduction for applications of inkjet-printed electronics. *Nanotechnology*, *19*(41), 415604.
26. Lu, A. H., Salabas, E. E., & Schüth, F. (2007). Magnetic nanoparticles: Synthesis, protection, functionalization, and application. *Angewandte Chemie International Edition*, *46*(8), 1222–1244.
27. Manikam, V. R., Cheong, K. Y., & Razak, K. A. (2011). Chemical reduction methods for synthesizing Ag and Al nanoparticles and their respective nanoalloys. *Materials Science and Engineering: B*, *176*(3), 187–203.
28. Marin, M. L., McGilvray, K. L., & Scaiano, J. C. (2008). Photochemical strategies for the synthesis of gold nanoparticles from Au (III) and Au (I) using photoinduced free radical generation. *Journal of the American Chemical Society*, *130*(49), 16572–16584.
29. Morozov, Y., Chistyakov, D., Chernyshev, V., & Sergeev, G. (2017). Cryochemical synthesis of polymorphous nanostructures of a steroid neurohormone. *Molecules*, *22*(8), 1378.
30. Nagasawa, H., Maruyama, M., Komatsu, T., Isoda, S., & Kobayashi, T. (2002). Physical characteristics of stabilized silver nanoparticles formed using a new thermal-decomposition method. *Physica Status Solidi (a)*, *191*(1), 67–76.
31. Olad, A., Alipour, M., & Nosrati, R. (2017). The use of biodegradable polymers for the stabilization of copper nanoparticles synthesized by chemical reduction method. *Bulletin of Materials Science*, *40*, 1013–1020.
32. Pastoriza-Santos, I., & Liz-Marzán, L. M. (1999). Formation and stabilization of silver nanoparticles through reduction by N, N-dimethylformamide. *Langmuir*, *15*(4), 948–951.
33. Peng, S., Wang, C., Xie, J., & Sun, S. (2006). Synthesis and stabilization of monodisperse Fe nanoparticles. *Journal of the American Chemical Society*, *128*(33), 10676–10677.
34. Pietrobon, B., & Kitaev, V. (2008). Photochemical synthesis of monodisperse size-controlled silver decahedral nanoparticles and their

remarkable optical properties. *Chemistry of Materials*, *20*(16), 5186–5190.

35. Pomogailo, A. D., Dzhardimalieva, G. I., Pomogailo, A. D., & Dzhardimalieva, G. I. (2014). Physical-chemical methods of nanocomposite synthesis. *Nanostructured Materials Preparation via Condensation Ways*, *5*(2), 91–139.
36. Rane, A. V., Kanny, K., Abitha, V. K., & Thomas, S. (2018). Methods for synthesis of nanoparticles and fabrication of nanocomposites. In *Synthesis of Inorganic Nanomaterials*. Woodhead Publishing. (Vol. 2, pp. 121–139).
37. Rath, M. C. (2021). Photo-and radiation-induced synthesis of nanomaterials. *Handbook on Synthesis Strategies for Advanced Materials: Volume-I: Techniques and Fundamentals*, *9*, 623–656.
38. Razi, M., Contreras-Mateus, M., Hashlamoun, K. W., & Nassar, N. N. (2022). Nanoparticles: Preparation, stabilization, and control over particle size. In *Nanoparticles: An Emerging Technology for Oil Production and Processing Applications*. Cham: Springer International Publishing. (Vol. 2, pp. 1–40).
39. Remita, H., & Lampre, I. (2024). Synthesis of metallic nanostructures using ionizing radiation and their applications. *Materials*, *17*(2), 364.
40. Rothgery, E. F., Holt, R. J., & McGee Jr, H. A. (1975). Cryochemical synthesis and molecular energetics of cyclopropanone and some related compounds. *Journal of the American Chemical Society*, *97*(17), 4971–4973.
41. Saha, B., Bhattacharjee, D., Sandilya, P., & Sengupta, S. (2019). Cryochemical approach to develop catalysts for intensification of the hydrodesulfurization reaction. *Industrial & Engineering Chemistry Research*, *58*(16), 6266–6277.
42. Saha, S., Pal, A., Kundu, S., Basu, S., & Pal, T. (2010). Photochemical green synthesis of calcium-alginate-stabilized Ag and Au nanoparticles and their catalytic application to 4-nitrophenol reduction. *Langmuir*, *26*(4), 2885–2893.
43. Sakamoto, M., & Majima, T. (2010). Photochemistry for the synthesis of noble metal nanoparticles. *Bulletin of the Chemical Society of Japan*, *83*(10), 1133–1154.
44. Sakamoto, M., Fujistuka, M., & Majima, T. (2009). Light as a construction tool of metal nanoparticles: Synthesis and mechanism. *Journal of Photochemistry and Photobiology C: Photochemistry Reviews*, *10*(1), 33–56.
45. Sathiyanarayanan, G., Dineshkumar, K., & Yang, Y. H. (2017). Microbial

3.1.1. Transmission Electron Microscopy

When subjected to a high-energy electron beam with an energy range of 50–200 keV, the thin film sample exhibits luminescence. This lighting takes place in a vacuum environment with a pressure of approximately 10–6 mm of Hg. Electrons are deflected at relatively small angles by atoms in the specimen and then allowed through the sample and are guided through a series of magnetic lenses to create a detailed image of the internal structure of the sample on a screen and film. An impressive level of precision is attained, with a resolution of 0.1 nm, equivalent to a magnification factor of 106 (Inkson, 2016). The resolution is determined by the unique characteristics of the sample and the methodology employed in its creation. Typically, films with a thickness of 0.01 micrometers are subjected to analysis. Using carbon copies can help to expand the range of contrast. With the help of modern ultramicrotomes, it is now possible to collect sections with an incredibly precise thickness ranging from 10 to 100 nanometers. This section primarily revolves around thin metal foils. Transmission microscopes enable the capture of diffraction patterns, which offer valuable insights into the crystal structure of a sample (Wang, 2003).

3.1.2. Scanning Electron Microscopy

This method is commonly employed for the analysis of surface particles. Using magnetic lenses, the size of a concentrated electron beam can be significantly reduced, resulting in a thin probe that measures between 1 to 10 millimeters in thickness. The investigation progresses systematically, moving methodically, across a material, thoroughly analyzing it with great efficiency. Electron-surface interaction results in the production of different emissions, including (Cazaux, 2005):

i. Secondary and reflected electrons,
ii. The transmitted electrons,
iii. X-ray slowing-down radiations,
iv. The optic radiation

It is possible to detect and convert all of the mentioned sources of radiation into electrical impulses. The signals are amplified before being transferred to a cathode-ray tube. A comparable scenario arises in TV kinescopes. Photography is a powerful tool for capturing and preserving images that can be displayed on a screen. One major benefit of this techni□ue is the extensive amount of information it offers, although a downside is the lengthy scanning periods it requires. High resolution is only possible when scanning speeds are kept low (Ul-Hamid, 2018). This method is often used for particles that are larger than Sn in size. There is a limitation on the thickness of the sample that exopolysaccharide-mediated synthesis and stabilization of metal nanoparticles. *Critical Reviews in Microbiology*, *43*(6), 731–752.

46. Sergeev, B. M., Kasaikin, V. A., Litmanovich, E. A., Sergeev, G. B., & Prusov, A. N. (1999). Cryochemical synthesis and properties of silver nanoparticle dispersions stabilized by poly (2-dimethylaminoethyl methacrylate). *Mendeleev Communications*, *9*(4), 130–132.
47. Sergeev, G. B., & Shabatina, T. I. (2008). Cryochemistry of nanometals. *Colloids and Surfaces A: Physicochemical and Engineering Aspects*, *313*, 18–22.
48. Shabatina, T. I., & Sergeev, G. B. (2010). Cryochemistry of nanometals. *Polymer Thin Films*, *185*, 6544–6548.
49. Shirokova, L. N., Revina, A. A., Aleksandrova, V. A., & Fenin, A. A. (2016). Radiation-chemical synthesis of silver nanoparticles in aqueous solution of chitin derivative. *Inorganic Materials: Applied Research*, *7*, 730–736.
50. Shlyakhtin, O., & Tretyakov, Y. (1999). Recent progress in cryochemical synthesis of oxide materials. *Journal of Materials Chemistry*, *9*(1), 19–24.
51. Singh, A., Mangla, P., Yumnam, S., Amgoth, C., & Rath, M. C. (2020). Radiation-induced synthesis of nanoparticles in aqueous solution: A mechanistic approach. In *AIP Conference Proceedings*. AIP Publishing. (Vol. 2269, pp. 98–100).
52. Spadaro, D., Barletta, E., Barreca, F., Curro, G., & Neri, F. (2010). Synthesis of PMA stabilized silver nanoparticles by chemical reduction process under a two-step UV irradiation. *Applied Surface Science*, *256*(12), 3812–3816.
53. Szczyglewska, P., Feliczak-Guzik, A., & Nowak, I. (2023). Nanotechnology–general aspects: A chemical reduction approach to the synthesis of nanoparticles. *Molecules*, *28*(13), 4932.
54. Thakkar, K. N., Mhatre, S. S., & Parikh, R. Y. (2010). Biological synthesis of metallic nanoparticles. *Nanomedicine: Nanotechnology, Biology and Medicine*, *6*(2), 257–262.
55. Toshima, N., & Yonezawa, T. (1998). Bimetallic nanoparticles—Novel materials for chemical and physical applications. *New Journal of Chemistry*, *22*(11), 1179–1201.
56. Trakhtenberg, L. I., & Gerasimov, G. N. (2005). Metal-containing polymers: Cryochemical synthesis, structure, and physicochemical properties. *Metal–Polymer Nanocomposites*, *37*, 546–550.
57. Trakhtenberg, L. I., Gerasimov, G. N., Aleksandrova, L. N., & Potapov, V.

K. (2002). Photo and radiation cryochemical synthesis of metal–Polymer films: Structure, sensor and catalytic properties. *Radiation Physics and Chemistry*, *65*(4–5), 479–485.

58. Trusova, E. A., & Trutnev, N. S. (2018). Cryochemical synthesis of ultrasmall, highly crystalline, nanostructured metal oxides and salts. *Beilstein Journal of Nanotechnology*, *9*(1), 1755–1763.
59. Verma, C., Ebenso, E. E., & Quraishi, M. A. (2019). Transition metal nanoparticles in ionic liquids: Synthesis and stabilization. *Journal of Molecular Liquids*, *276*, 826–849.
60. Vernaya, O. I., Shabatin, V. P., Nuzhdina, A. V., Zvukova, N. D., Khvatov, D. I., Semenov, A. M., & Melnikov, M. Y. (2017). Cryochemical synthesis and antibacterial activity of hybrid nanocomposites based on dioxidine containing Ag and Cu nanoparticles incorporated in biopolymer cryostructurates. *Russian Chemical Bulletin*, *66*, 2152–2156.
61. Virkutyte, J., & Varma, R. S. (2011). Green synthesis of metal nanoparticles: Biodegradable polymers and enzymes in stabilization and surface functionalization. *Chemical Science*, *2*(5), 837–846.
62. Vollmer, C., & Janiak, C. (2011). Naked metal nanoparticles from metal carbonyls in ionic liquids: Easy synthesis and stabilization. *Coordination Chemistry Reviews*, *255*(17–18), 2039–2057.
63. Weare, W. W., Reed, S. M., Warner, M. G., & Hutchison, J. E. (2000). Improved synthesis of small (d core≈ 1.5 nm) phosphine-stabilized gold nanoparticles. *Journal of the American Chemical Society*, *122*(51), 12890–12891.
64. Yelenich, O. V., Solopan, S. O., Kolodiazhnyi, T. V., Greneche, J. M., & Belous, A. G. (2015). Synthesis of iron oxide nanoparticles by different methods and study of their properties. *Solid State Phenomena*, *230*, 108–113.
65. Zezin, A. A., Klimov, D. I., Zezina, E. A., Mkrtchyan, K. V., & Feldman, V. I. (2020). Controlled radiation-chemical synthesis of metal polymer nanocomposites in the films of interpolyelectrolyte complexes: Principles, prospects and implications. *Radiation Physics and Chemistry*, *169*, 108076.
66. Zharikov, A. A., Zezin, A. A., Zezina, E. A., Emel'yanov, A. I., & Prozorova, G. F. (2020). The radiational chemical reduction of silver ions in water suspensions of poly (vinyltriazole): The kinetic peculiarities of the formation of nanoparticles. *Moscow University Chemistry Bulletin*, *75*, 87–91.

CHAPTER 3

CHARACTERIZATION TECHNIQUES FOR NANOMATERIALS

LEARNING OBJECTIVES

At the end of this chapter, readers will be able to:

1. Understand the principles of electron microscopy.
2. Comprehend the principles and applications of probe microscopy.
3. Gain knowledge about diffraction techniques for crystallographic analysis.
4. Comprehend miscellaneous techniques such as EXAFS, x-ray fluorescence spectroscopy, photoelectron spectroscopy, and mass spectrometry.

INTRODUCTORY EXAMPLE

Nanosys, Inc. is a well-known company that specializes in nanotechnology. They specialize in the development and commercialization of quantum dot technology for display applications. They employ a range of cutting-edge characterization techniques to ensure the exceptional quality and performance of nanomaterials. Understanding the properties of quantum dots at the nanoscale, improving their effectiveness, and maintaining high standards in product development require the use of these methodologies.

Nanosys utilize Transmission Electron Microscopy (TEM) to capture high-resolution images of quantum dots. This advanced technique allows to thoroughly analyze the size, shape, and internal structure of these nanoparticles. Utilizing Scanning Electron Microscopy (SEM) elevates this process by providing detailed visuals of the surface structure, assisting the corporation in ensuring a high level of consistency and excellence in their □uantum dot films.

At Nanosys, X-ray Diffraction (XRD) is a widely utilized technique for analyzing the crystal structure and phase composition of quantum dots. With a deep understanding of the crystallographic features of quantum dots, Nanosys has the ability to tailor the production process and achieve specific optical characteristics.

Nanosys ensures that its quantum dots adhere to the strict quality standards required for top-notch display technologies through the implementation of these characterization methodologies.

UNIT INTRODUCTION

The size and physicochemical properties of nanoparticles are closely connected and have a significant impact on the investigation of their chemical transformations. Furthermore, numerous approaches are available for examining the properties of particles situated on the surface as well as within the bulk (Lin et al., 2014).

The primary methods used to determine the size and specific properties of nanoparticles in the gas phase are as follows:

i. The process of ionization using photons and electrons, followed by the examination of the resulting mass spectra using quadrupole and time-of-flight mass spectrometers.

ii. The process of breaking down atoms into neutral clusters and then selecting them based on their masses.

iii. On grids use of electron transmission microscopy to gather information about the structure and size of particles (Kumar et al., 2019).

The following procedures are employed to acquire details about the particles situated on the surface:

i. Transmission and scanning electron microscopies (TEM, SEM) yield data regarding the dimensions, morphology, dispersion, and surface characteristics of particles.

ii. Electron diffraction yields valuable data regarding the dimensions, arrangement, states, and distances between atoms in particles.

iii. Scanning tunneling microscopy (STM) is employed to ascertain the dimensions, morphology, and interior composition of particles.

iv. The adsorption of gasses yields data regarding the surface area of particles.

v. Photoelectron spectroscopy is employed to ascertain the electronic configuration of particles.

vi. Conductivity measurements yield data regarding the conduction band, percolation, and topology of particles (Prasad et al., 2024).

Various methods are also employed to ascertain the dimensions and specific characteristics of nanoparticles either in a large quantity or when embedded in a material. The techniques of TEM, SEM, conductivity tests, and electron diffraction provide data on particles in the bulk, which is similar to the information gained for particles on the outer layer. Various other methodologies are employed to investigate particles in their entirety. X-ray diffraction is a useful technique for determining both the sizes of particles and the structures inside them (Mudalige et al., 2019).

Using the EXAFS method, one can determine particle sizes, while EPR and NMR provide valuable insights into the electronic structure. Mossbauer spectroscopy is a fascinating technique that involves the absorption of gamma rays by atomic nuclei in materials, known as gamma resonance. It is often employed to analyze the internal structure of different substances, such as iron (Salame et al., 2018). Gamma quanta possess a rather moderate energy level of approximately 150 kilo electron volts (keV), and upon absorption, they trigger a reaction within the nucleus. When the excitation energy of the nucleus aligns with the energy difference between the internal energy of the nucleus in the excited and ground states during a quantum transition, the resonance condition occurs. Understanding the energy shift relies on the characteristics of a nucleus and provides valuable knowledge about the intricate composition of materials.

This approach has limitations, but it offers valuable insights for Fe, Sn, and Te (Boddolla & Thodeti, 2018) (Figure 3.1).

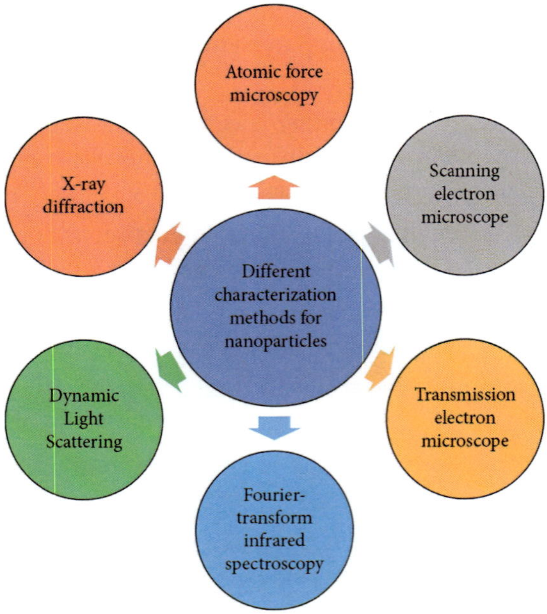

Figure 3.1. *Illustration of different characterization techniques for nanoparticles.*

Source: Apoorva Sharma, Creative Commons License.

3.1. ELECTRON MICROSCOPY

Learning Objectives

- Understand the operational principles and applications of transmission electron microscope
- Comprehend the principles and applications of scanning electron microscope

This discussion will thoroughly delve into the topic of microscopy, which plays a crucial role in determining the size of nanoparticles. Usually, this relates to the realm of electron microscopy, where accelerated electron beams and different types of probe microscopes are used. There are two main branches of electron microscopy (Zhang & Su, 2014):

i. TEM
ii. SEM

restricts the utilization of this approach. For electrons with an energy of 100 kilo electron volts, a sample thickness of around 50 nanometers is required. Various techniques are utilized to ensure proper sample preparation, minimizing the risk of damage or loss. In addition, it is important to consider the possible effects of electron emission on the samples, such as the formation of particle clusters caused by the electron beam (Mohammed & Abdullah, 2018).

One commonly employed technique for sample preparation involves the use of ultramicrotomes. However, their effectiveness is limited when dealing with unevenly distributed deposits, especially in the case of islets. Matrix dissolving is a frequently used technique in chemical methods. Understanding the meaning of a histogram derived from microscopic examinations can be influenced by the techni□ue employed to create the specimen. During the 1980s, significant advancements were made in the field of electron microscopy. Energy-loss spectrometers were utilized to develop microscopes capable of analyzing elemental composition through computer technology (Hafner, 2007). Energy-loss spectrometry was employed in conjunction with transmission electron microscopy (TEM) and scanning electron microscopy (SEM). By reconfiguring the magnetic-prism system, it becomes possible to manipulate the contrast of the image. Various factors can affect the contrast, including the angle at which light hits the surface, the atomic number of the material, and its reflective properties. With the help of cutting-edge technologies, scientists can now capture incredibly detailed images of different elements, ranging from boron to uranium. These images have an impressive spatial resolution of 0.5 nm and a remarkable sensitivity of up to 10–20 g. To put it into perspective, this level of sensitivity is equivalent to detecting approximately 150 atoms of calcium (Zhou et al., 2007).

The advent of computer systems for image processing played a pivotal role in advancing electron microscopy. These strategies made it possible to gather histograms that are related to forms, orientations, and sizes. At present, it is feasible to dissect individual components of the framework, conduct statistical analysis of data, estimate localized microconcentrations, and ascertain lattice features. With integrated processors, users can enjoy a wide range of control options and achieve comprehensive control over microscopes (Leamy, 1982) (Figure 3.2).

Did You Know?

In 1931, a remarkable achievement was made with the creation of the first practical electron microscope, surpassing the resolution capabilities of the light microscopes of its era.

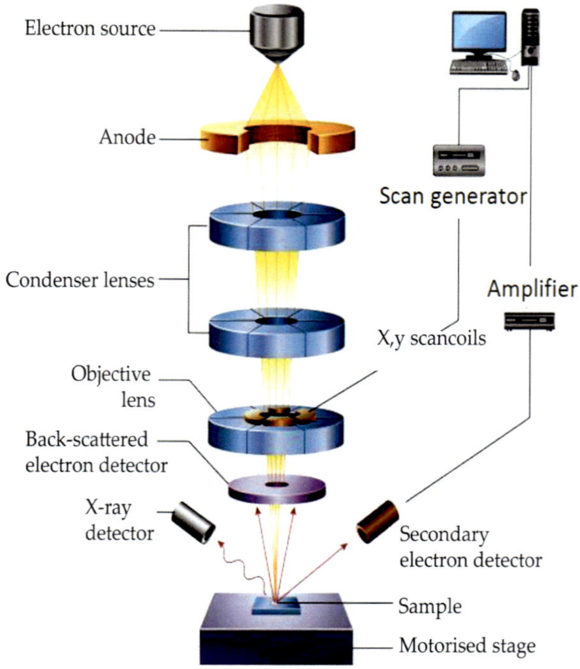

Figure 3.2. *Schematic of scanning electron microscope.*

Source: D.B. Williams, Creative Commons License.

3.2. PROBE MICROSCOPY

Learning Objectives

- Comprehend the fundamentals of Atomic Force Microscopy (AFM) focusing on its mechanisms and specific applications in analyzing nanomaterials

In Scanning Tunneling Microscope, a probe is used in these microscopes to make contact with the surface, either mechanically or through tunneling. This action leads to the achievement of a state of balance between the interactions of the probe and the sample. The equilibrium at hand encompasses the interplay of various forces, both attractive and repulsive (Bottomley et al., 1996).

Scanning commences once equilibrium is achieved. The probe methodically scans the surface, taking into account the number of line scans, their length, and the distance between each line. The probe is operated by a piezomanipulator that changes in size due to the applied voltage differential. The motion of a specimen in three dimensions is illustrated in Figure 3.3 (Bian et al., 2021).

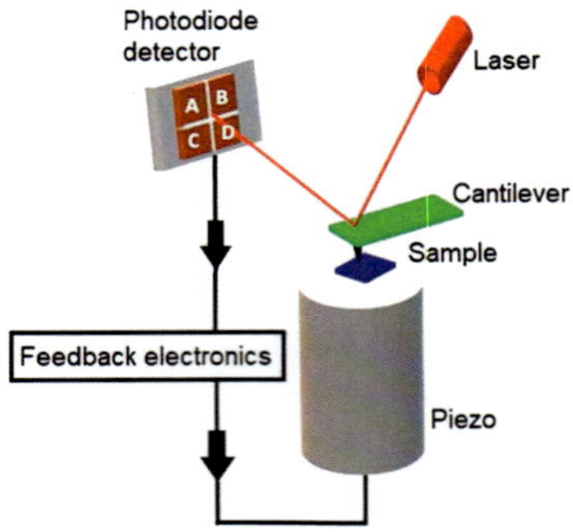

Figure 3.3. *Schematic of the operation of AFM mechanism.*

Source: Morris, Creative Commons License.

During scanning, the feedback mechanism fine-tunes the probe's position in the Z direction to maintain a steady signal level that matches the intended interaction amplitude. This interaction is denoted as A(X, Y, Z). The computer optimizes the signals on plates X, Y, and Z of the piezomanipulator. During the scanning process, there is a correlation between the changes in Z and the relationship between A and (X, Y) (Fischer et al., 2016). This connection is characterized by the consistent value of A and the distinct topographical characteristics of the surface under investigation. The surface topography resulting from ongoing interaction A can be described as a type of interdependence. As the strength of interaction A(X, Y, Z) varies across the surface, the complexity of the resulting image becomes more pronounced. The display presents a blend of maps illustrating the surface topography and the spatial distribution of A intensity (Horcas et al., 2007).

Scanning probe microscopy enables the gathering of different dependencies beyond surface texture under constant-interaction conditions. The e□uation Φ |A = const (X, Y) provides valuable insights into different surface characteristics. The function Φ (X, Y) is calculated using a constant value for the interaction A. Estimating a consistent distance between the probe and the sample is one approach to achieve this. The function Φ (X, Y) can □uantify the level of interactions that are not related to A (Poggi et al., 2004).

The variation in Φ (X, Y) is commonly used in different probe microscopes, while condition A remains constant. When studying the interaction between the probe and sample in these microscopes, a precise distance is always maintained. This is accomplished through closed-contact and patched-contact modes (Greene et al., 2004).

The circuit depicted in Figure 3.3 is employed in atomic force microscopes to precisely control and maintain the force interaction at a specific level. The probe is located at the free and unsecured end of a flexible arm cantilever. When the probe comes close to or makes contact with the sample, their interaction results in the bending of the cantilever. This bending is measured by a highly precise transducer (Rao & Biswas, 2009). It is essential to maintain the stability of the curvature at the desired magnitude during scanning to accurately reproduce the surface profile.

The level of engagement depends on the extent of the curvature. Atoms are typically represented as hemispheres in visualizations. Optical sensors are commonly found in most atomic force microscopy (AFM) devices. When a laser beam strikes the lever surface at an oblique angle, it is redirected toward the central region of a photodiode that is divided into four segments (Zhong & Yan, 2016). During nighttime, the lever's presence causes interference with the signals detected by the corresponding segments of the photodiode. The computer screen displays the signals representing the friction forces experienced during scanning, which are transmitted from the left and right segments. A thorough examination is performed on the signals emanating from the upper and lower segments, utilizing proportional and integrative feedback circuits (Fujita et al., 2009).

AFM allows for the use of a tailored contact mode. To accomplish this goal, a supplementary piezomanipulator is used to generate precise oscillations of the lever. The dynamics of the interaction between the probe and the sample are complex, involving their physical characteristics and the actions they take. These factors are determined by the properties of the probe and sample materials, as well as the parameters of the surrounding media. Studying polyparaxylylene films presents a significant challenge in terms of probe adhesion to the films (Angeloni et al., 2018).

The probe applies significant pressure, up to 109Pa, which could potentially exceed the maximum tolerance of different materials. Mica and graphite are frequently utilized as substrates. These materials have impressive scalability and boast surfaces that are flawlessly polished. AFM allows for the performance of nondestructive testing (Yamamoto et al., 2009). This phenomenon can be understood by considering the dispersion of pressure in multiple directions and the short duration of its effectiveness, which only lasts for about 10–5 seconds. Researchers are currently focused on reducing the forces of interaction between the probe and the sample, which remains a significant challenge. One possible strategy is to meticulously choose suitable materials for the study and utilize a range of contact methods (Akita & Nakayama, 2002).

Practice Problem

As a researcher, studying the surface topography of a nanomaterial at atomic resolution is of great interest. Which technique would be most suitable, and what factors are crucial in determining the resolution of this technique?

Solution to Practice Problem

Using Atomic Force Microscopy (AFM) would be highly beneficial for the researcher.

Factors that have a significant impact on AFM resolution:

i. Tip Sharpness: A higher resolution can be achieved with a sharper AFM tip.
ii. Minimize the forces between the tip and sample to prevent any distortion in their interaction.
iii. Vibration Isolation: Minimize external vibrations to improve image clarity.
iv. A slower scan rate generally enhances resolution.
v. Performing AFM in a controlled environment is essential to minimize thermal drift and contamination.

3.3. DIFFRACTION TECHNIQUES

Learning Objectives

- Understand the principles of X-ray diffraction technique
- Comprehend the mechanism of neturon diffraction technique to determine crystal structure and lattice parameters of nanomaterials.

X-ray and neutron diffraction techniques are highly specialized compared to electron microscopy. Furthermore, the examination of diffraction patterns that arise from the atomic structures of individual particles can be employed to investigate extremely tiny particles (Giannini et al., 2016). The reflex angle's width, denoted as Δ , shows a direct correlation with the increase in particle size (Δ ~1/R), known as the Scherrer effect. Smaller sizes are linked to a reduction in the number of lattice planes causing interference at the diffraction spot, whereas larger clusters typically exhibit diffraction rings (Bishnoi et al., 2017).

3.3.1. X-Ray Diffraction

When metal particles, crystals, and molecules interact with X-rays, they undergo a process called scattering. The incident beam of radiation generates secondary rays that have the same wavelength, typically ranging from 0.5 to 5 angstroms. The paths and strengths of these additional rays are determined by the makeup of the scattering samples. The size and arrangement of particles also have an impact on the magnitude of a diffracted ray. Polycrystalline particles generate secondary ray cones, with each cone corresponding to a specific set of crystal planes. When dealing with small and abundant crystals, it is observed that the cone maintains its structure, leading to the formation of a ring that appears darkened and irregular (Xiao et al., 2005).

Crystals are fascinating natural objects that possess a unique ability to diffract light, resulting in a beautifully organized pattern. For optimal research purposes, the crystals used must possess dimensions of approximately 0.1 mm and exhibit flawless architecture. When studying the arrangement of a moderately complex structure, which consists of 50–100 atoms in a single unit cell, it is common to observe multiple diffraction reflections (Moroz, 2011). These reflections can sometimes extend into the hundreds or even thousands. This technique is carried out using computer-controlled microdensitometers and diffractometers. In the past, these treatments used to take several months to complete, but now they can be accomplished within just 24 hours (Sharma et al., 2012).

Did You Know?

XRD played a pivotal role in the discovery of the DNA double helix structure by Watson and Crick.

X-ray diffraction is a valuable method for investigating amorphous materials and partially ordered particles in polycrystals. It enables the analysis of the phase composition, size, and alignment of grains, commonly referred to as texture. Low-angle scattering is a commonly employed technique for investigating spatial heterogeneities within a size range of 5–10Å. This method is used to analyze porous materials and finely dispersed alloys (Lamas et al., 2017).

Understanding atomic structure requires analyzing dimensions, and geometry, and classifying it into a symmetry group. This method requires a detailed analysis and involves quantitatively adjusting the size of all diffraction rays. Studying materials through X-ray diffraction methods allows for the exploration of various aspects such as equilibrium and nonequilibrium phases, substance compositions, residual stresses, phase diagrams, and related phenomena (Baer et al., 2008) (Figure 3.4).

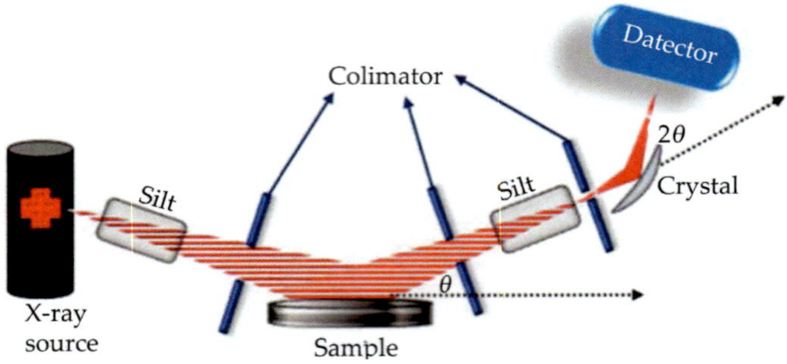

Figure 3.4. *Illustration of X-ray diffraction technique.*

Source: William Stillwell, Creative Commons License.

3.3.2. Neutron Diffraction

Neutrons have unique properties that make them ideal for studying a wide range of materials. Thermal neutrons, generated by nuclear reactors, have a maximum energy of 0.06eV. This energy level corresponds to the de Broglie wave with a wavelength of approximately 1Å, which coincides with the interatomic distances. This serves as the foundation for the methodology employed in structural neutron diffraction. Neutron spectroscopy takes advantage of the similarity in energy levels between thermal neutrons and thermal atomic oscillations. On the other hand, magnetic neutron diffraction depends on the existence of a magnetic moment (Petrunin, 2014).

3.4. MISCELLANEOUS TECHNIQUES

Learning Objectives

- Comprehend the application of EXAFS in determining local atomic environments
- Understand how XRF is used for elemental analysis and □uantification
- Gain insight about the mass spectrometry technique
- Comprehend the utility of photoelectron spectroscopy in assessing elemental composition, chemical states, and electronic structures

3.4.1. EXAFS

The technique relies on performing measurements at the absorption edge, particularly the K shell. In this scenario, the oscillations observed are a result of the interference between primary waves and secondary waves that are scattered by neighboring atoms. The energy of the photons plays a crucial role in this phenomenon (Chevrier et al., 2015). Through careful analysis of these oscillations, one can uncover valuable insights about the distances between adjacent atoms and observe any deviations in the surface layer of particles, in contrast to tightly packed metal lattices. Grasping the optical properties of metallic particles requires a thorough understanding of these fluctuations. These variations can impact the size of particles, the abundance of electrons, and the optical properties (Jajko et al., 2021).

3.4.2. X-Ray Fluorescence Spectroscopy

For this method, a low-intensity X-ray source is used to produce X-rays, which are then used to activate atoms within the substance that is being investigated. A secondary fluorescence radiation is produced as a consequence of this, and it is directed in the direction of the crystal analyzer. The analyzer first reflects the emission, and then a detector that is compatible with the analyzer detects the emission. While the goniometer is in charge of managing the detector, it is also responsible for controlling the crystal analyzer (Yıgıt et al., 2009). There is a certain wavelength of secondary emission that corresponds to each fixed point on the goniometer, and the analyzer chooses this wavelength with great care. In a sample, the intensities of the spectral lines have a direct and strong link with the quantitative amounts of components that are present. This relationship is straightforward and strong. When determining concentrations, it is common practice to compare the values obtained from reference samples to the acquired concentrations. Through the utilization of X-ray fluorescence technology, it is possible to conduct a speedy and precise study of the numerous elements that are listed in the Periodic Table. Many other types of samples can be analyzed with this approach, such as liquids, solids, powders, and films (Moradi et al., 2015).

Remember

XRF is a non-invasive technique that allows for the analysis of valuable artifacts and artworks, providing valuable insights into their composition without any risk of damage.

3.4.3. Mass Spectrometry

This technique is used to differentiate ionized species based on how they interact with electromagnetic fields and their masses. Ions' transit time over a given distance is measured via dynamic devices. Using a transverse electric field produced by a quadrupole capacitor—a device with four symmetric rods arranged around its central axis—the ions are separated in a quadrupole mass spectrometer. With a time-of-flight mass spectrometer, an ion cluster is inserted into the analyzer via a grid and moves in the direction of a collector inside the analyzer. During this process, the lighter ions (m2) move more quickly than the heavier ions (m1). The reason for the division of the ion packet is that although all of the ions in the original packet had the same amount of energy, their flight times and speeds along the analyzer t were inversely proportional to the square root of their masses. The connection between the analyzer length (L), the accelerating potential (v), and the time of flight (t) is shown by the equation t = L(m/2v)1/2 (Yagnik et al., 2016).

3.4.4. Photoelectron Spectroscopy

This technique relies on measuring the energy of electrons emitted from solids when they are exposed to light. Based on Einstein's theory, the total energy of a free electron, along with its kinetic energy, is equivalent to the energy of the photon that triggered its liberation. The acquired spectra are utilized to ascertain the binding energies and energy levels of electrons in the material under investigation. This method enables the examination of electron dispersion in the conduction bands, along with investigating material composition and the nature of chemical bonding. Metals possess a remarkable capacity to reflect photons and exhibit a notable interaction with electrons in the conduction band. The quantum yield has a relatively low value of e/1 photon, which contributes to the challenges associated with implementing this technology (Gengenbach et al., 2021).

SUMMARY

- This chapter highlights key learning aspects for readers, with a specific focus on the various approaches used to identify nanomaterials and their uni□ue contributions to the field.
- This chapter examines the basic principles and practical applications of Transmission Electron Microscopy (TEM) and Scanning Electron Microscopy (SEM). TEM provides intricate atomic-level pictures that are crucial for evaluating the morphology and crystallography of nanomaterials. SEM offers detailed visualizations of surface topography and composition, which are essential for examining surface structure and identifying surface constituents.
- The chapter explains the use of Atomic Force Microscopy (AFM). AFM is well-known for its ability to provide three-dimensional topographical data and measure surface roughness, mechanical properties, and electrical characteristics.
- The chapter also provides a clear explanation of X-ray Diffraction and Neutron Diffraction. X-ray diffraction is essential for determining the crystal structure, phase composition, and lattice properties of crystalline materials. Neutron diffraction, similar to XRD, is particularly beneficial for analyzing materials that have light atoms, magnetic properties, and complex crystal structures.
- The chapter ends by examining supplementary methods such as X-ray Fluorescence Spectroscopy (XRF), Extended X-ray Absorption Fine Structure (EXAFS), Mass Spectrometry, and Photoelectron Spectroscopy (PES).

MULTIPLE CHOICE QUESTIONS

1. Which of the following techniques provides atomic-level images of nanomaterials by transmitting electrons through a thin specimen?

a. Scanning Electron Microscopy (SEM)

b. Atomic Force Microscopy (AFM)

c. Transmission Electron Microscopy (TEM)

d. X-ray Diffraction (XRD)

2. What does Atomic Force Microscopy (AFM) primarily measure?

a. Surface composition

b. Three-dimensional topography

c. Crystal structure

d. Elemental composition

3. Which technique uses a conductive tip to scan the surface of a material at a close distance to provide information on surface electronic states?

a. Scanning Electron Microscopy (SEM)

b. Scanning Tunneling Microscopy (STM)

c. X-ray Photoelectron Spectroscopy (XPS)

d. Extended X-ray Absorption Fine Structure (EXAFS)

4. X-ray Diffraction (XRD) is primarily used to determine which of the following properties of crystalline materials?

a. Surface roughness

b. Phase composition and lattice parameters

c. Elemental composition

d. Electrical characteristics

5. Which characterization technique is especially useful for studying materials with light atoms and magnetic properties?

a. Neutron Diffraction

b. X-ray Diffraction (XRD)

c. Mass Spectrometry (MS)

d. X-ray Fluorescence Spectroscopy (XRF)

REVIEW QUESTIONS

1. What are the primary differences between Transmission Electron Microscopy (TEM) and Scanning Electron Microscopy (SEM)?
2. How does X-ray Diffraction (XRD) help in determining the crystal structure of nanomaterials?
3. Why is Neutron Diffraction particularly useful for studying materials with light atoms and magnetic properties?
4. What is the principle behind X-ray Fluorescence Spectroscopy (XRF), and how is it used in the elemental analysis of nanomaterials?
5. Why is it important to use a combination of different characterization techniques when studying nanomaterials?

Answers to Multiple Questions

1. (c); 2. (b); 3. (b); 4. (b); 5. (a)

REFERENCES

1. Akita, S., & Nakayama, Y. (2002). Manipulation of nanomaterial by carbon nanotube nanotweezers in scanning probe microscope. *Japanese Journal of Applied Physics*, *41*(6S), 4242.
2. Angeloni, L., Reggente, M., Passeri, D., Natali, M., & Rossi, M. (2018). Identification of nanoparticles and nanosystems in biological matrices with scanning probe microscopy. *Wiley Interdisciplinary Reviews: Nanomedicine and Nanobiotechnology*, *10*(6), 1521.
3. Baer, D. R., Amonette, J. E., Engelhard, M. H., Gaspar, D. J., Karakoti, A. S., Kuchibhatla, S., & Wang, C. M. (2008). Characterization challenges for nanomaterials. *Surface and Interface Analysis: An International Journal Devoted to the Development and Application of Techniques for the Analysis of Surfaces, Interfaces and Thin Films*, *40*(3–4), 529–537.
4. Bian, K., Gerber, C., Heinrich, A. J., Müller, D. J., Scheuring, S., & Jiang, Y. (2021). Scanning probe microscopy. *Nature Reviews Methods Primers*, *1*(1), 36.
5. Bishnoi, A., Kumar, S., & Joshi, N. (2017). Wide-angle X-ray diffraction (WXRD): Technique for characterization of nanomaterials and polymer nanocomposites. In *Microscopy Methods in Nanomaterials Characterization*. Elsevier. (Vol. 1, pp. 313–337).
6. Boddolla, S., & Thodeti, S. (2018). A review on characterization techniques of nanomaterials. *International Journal of Engineering, Science and Mathematics*, *7*(1), 169–175.
7. Bottomley, L. A., Coury, J. E., & First, P. N. (1996). Scanning probe microscopy. *Analytical Chemistry*, *68*(12), 185–230.
8. Cazaux, J. (2005). Recent developments and new strategies in scanning electron microscopy. *Journal of Microscopy*, *217*(1), 16–35.
9. Chevrier, D. M., Yang, R., Chatt, A., & Zhang, P. (2015). Bonding properties of thiolate-protected gold nanoclusters and structural analogs from X-ray absorption spectroscopy. *Nanotechnology Reviews*, *4*(2), 193–206.
10. Fischer, M. C., Wilson, J. W., Robles, F. E., & Warren, W. S. (2016). Invited review article: Pump-probe microscopy. *Review of Scientific Instruments*, *87*(3), 852–860.
11. Fujita, D., Onishi, K., & Xu, M. (2009). Standardization of nanomaterials characterization by scanning probe microscopy for societal acceptance. In *Journal of Physics: Conference Series*. IOP Publishing. (Vol. 159, pp. 12002–12005).

12. Gengenbach, T. R., Major, G. H., Linford, M. R., & Easton, C. D. (2021). Practical guides for x-ray photoelectron spectroscopy (XPS): Interpreting the carbon 1s spectrum. *Journal of Vacuum Science & Technology A*, *39*(1), 123–126.
13. Giannini, C., Ladisa, M., Altamura, D., Siliqi, D., Sibillano, T., & De Caro, L. (2016). X-ray diffraction: A powerful technique for the multiple-length-scale structural analysis of nanomaterials. *Crystals*, *6*(8), 87.
14. Greene, M. E., Kinser, C. R., Kramer, D. E., Pingree, L. S., & Hersam, M. C. (2004). Application of scanning probe microscopy to the characterization and fabrication of hybrid nanomaterials. *Microscopy Research and Technique*, *64*(5–6), 415–434.
15. Hafner, B. (2007). Scanning electron microscopy primer. *Characterization Facility, University of Minnesota-Twin Cities*, *1*(2), 1–29.
16. Horcas, I., Fernández, R., Gomez-Rodriguez, J. M., Colchero, J., Gómez-Herrero, J., & Baro, A. M. (2007). WSXM: A software for scanning probe microscopy and a tool for nanotechnology. *Review of Scientific Instruments*, *78*(1), 52–56.
17. Inkson, B. J. (2016). Scanning electron microscopy (SEM) and transmission electron microscopy (TEM) for materials characterization. In *Materials Characterization Using Nondestructive Evaluation (NDE) Methods*. Woodhead Publishing. (Vol. 2, pp. 17–43).
18. Jajko, G., Kozyra, P., Strzempek, M., Indyka, P., Zając, M., Witkowski, S., & Piskorz, W. (2021). Structural studies of aluminated form of zeolites—EXAFS and XRD experiment, STEM micrography, and DFT modelling. *Molecules*, *26*(12), 3566.
19. Kumar, P. S., Pavithra, K. G., & Naushad, M. (2019). Characterization techniques for nanomaterials. In *Nanomaterials for Solar Cell Applications*. Elsevier. (Vol. 3, pp. 97–124).
20. Lamas, D. G., de Oliveira Neto, M., Kellermann, G., & Craievich, A. F. (2017). X-ray diffraction and scattering by nanomaterials. In *Nanocharacterization Techniques*. William Andrew Publishing. (Vol. 2, pp. 111–182).
21. Leamy, H. J. (1982). Charge collection scanning electron microscopy. *Journal of Applied Physics*, *53*(6), 51–80.
22. Lin, P. C., Lin, S., Wang, P. C., & Sridhar, R. (2014). Techniques for physicochemical characterization of nanomaterials. *Biotechnology Advances*, *32*(4), 711–726.
23. Mohammed, A., & Abdullah, A. (2018). Scanning electron microscopy

(SEM): A review. In *Proceedings of the 2018 International Conference on Hydraulics and Pneumatics—HERVEX*, Băile Govora, Romania (Vol. 218, pp. 7–9).

24. Moradi, M., Yamini, Y., Kakehmam, J., & Ahmadi, K. (2015). A review in the sample preparation of aqueous solutions combined with X-ray fluorescence detection. *Journal of the Iranian Chemical Society*, *12*, 831–838.
25. Moroz, E. M. (2011). X-ray diffraction structure diagnostics of nanomaterials. *Russian Chemical Reviews*, *80*(4), 293.
26. Mudalige, T., Qu, H., Van Haute, D., Ansar, S. M., Paredes, A., & Ingle, T. (2019). Characterization of nanomaterials: Tools and challenges. *Nanomaterials for Food Applications*, *4*(3), 313–353.
27. Petrunin, V. F. (2014). Neutron diffraction investigation of specific features of the structure of ultrafine (nano) materials. *Physics of the Solid State*, *56*, 171–177.
28. Poggi, M. A., Gadsby, E. D., Bottomley, L. A., King, W. P., Oroudjev, E., & Hansma, H. (2004). Scanning probe microscopy. *Analytical Chemistry*, *76*(12), 3429–3444.
29. Prasad, R. D., Prasad, R. S., Prasad, R. B., Prasad, S. R., Singha, S. B., Singha, D., & Navathe, G. J. (2024). A review on modern characterization techniques for analysis of nanomaterials and biomaterials. *ES Energy & Environment*, *23*, 1087.
30. Rao, C. N. R., & Biswas, K. (2009). Characterization of nanomaterials by physical methods. *Annual Review of Analytical Chemistry*, *2*, 435–462.
31. Salame, P. H., Pawade, V. B., & Bhanvase, B. A. (2018). Characterization tools and techniques for nanomaterials. In *Nanomaterials for Green Energy*. Elsevier. (Vol. 2, pp. 83–111).
32. Sharma, R., Bisen, D. P., Shukla, U., & Sharma, B. G. (2012). X-ray diffraction: A powerful method of characterizing nanomaterials. *Recent Research in Science and Technology*, *4*(8), 77–79.
33. Ul-Hamid, A. (2018). A beginners' guide to scanning electron microscopy. Cham, Switzerland: Springer International Publishing. (Vol. 1, pp. 402–405).
34. Wang, Z. L. (2003). New developments in transmission electron microscopy for nanotechnology. *Advanced Materials*, *15*(18), 1497–1514.
35. Xiao, Y., Cai, Z., Wang, Z. L., Lai, B., & Chu, Y. S. (2005). An X-ray nano diffraction technique for structural characterization of individual nanomaterials. *Journal of Synchrotron Radiation*, *12*(2), 124–128.

36. Yagnik, G. B., Hansen, R. L., Korte, A. R., Reichert, M. D., Vela, J., & Lee, Y. J. (2016). Large scale nanoparticle screening for small molecule analysis in laser desorption ionization mass spectrometry. *Analytical Chemistry*, *88*(18), 8926–8930.
37. Yamamoto, T., Murakami, Y., Motoyanagi, J., Fukushima, T., Maruyama, S., & Kato, M. (2009). An analytical system for single nanomaterials: Combination of capillary electrophoresis with Raman spectroscopy or with scanning probe microscopy for individual single-walled carbon nanotube analysis. *Analytical Chemistry*, *81*(17), 7336–7341.
38. Yıgıt, D., Baydas, E., & Güleryüz, M. (2009). Elemental analysis of various cherry fruits by wavelength dispersive X-ray fluorescence spectrometry. *Asian Journal of Chemistry*, *21*(4), 2935–2942.
39. Zhang, B., & Su, D. S. (2014). Transmission electron microscopy and the science of carbon nanomaterials. *Small*, *10*(2), 222–229.
40. Zhong, J., & Yan, J. (2016). Seeing is believing: Atomic force microscopy imaging for nanomaterial research. *RSC Advances*, *6*(2), 1103–1121.
41. Zhou, W., Apkarian, R., Wang, Z. L., & Joy, D. (2007). Fundamentals of scanning electron microscopy (SEM). In *Scanning Microscopy for Nanotechnology: Techniques and Applications*, *5*(2), 1–40.

CHAPTER 4

NANOTECHNOLOGY AND HETEROGENEOUS CATALYSIS

LEARNING OBJECTIVES

At the end of this chapter, readers will be able to:

1. Understand how nanotechnology enhances the synthesis and performance of catalysts.
2. Learn about nanoreactors and their significance in heterogeneous catalysis.
3. Comprehend the importance of catalytic processes in the synthesis of nanomaterials, such as nanotubes.

INTRODUCTORY EXAMPLE

BASF, a global leader in the chemical industry, has been at the forefront of incorporating nanotechnology into heterogeneous catalysis to enhance chemical processes and product performance. One significant development by BASF is their innovative use of nanocatalysts in the production of clean fuels.

BASF has developed advanced nanocatalysts that play a critical role in refining processes, particularly in the production of ultra-low sulfur diesel (ULSD). These nanocatalysts are designed to provide higher activity and selectivity, which leads to more efficient removal of sulfur compounds from diesel fuel. The use of nanotechnology allows BASF to create catalysts with precisely controlled particle sizes and surface structures, optimizing the active sites for enhanced performance.

UNIT INTRODUCTION

In the last decade, nanotechnology and nanoscience have gained significant popularity as areas of research and development on a global scale. Heterogeneous catalysis was mentioned as an illustrative example to highlight its significant potential and the numerous advantages it offers to society. Therefore, it is logical to anticipate that the rapid growth in advancements in nanotechnology and nanoscience will have a substantial influence on the comprehension, implementation, and utilization of catalysis. This chapter provides a concise overview of how recent advancements in heterogeneous catalysis have contributed to improved control of catalytic processes and the application of catalysis in nanotechnology. The discussion is supported by selected examples (Kung & Kung, 2004).

Nanotechnology, as defined by the U.S. National Nanotechnology Institute, focuses on the scientific exploration and manipulation of structures, devices, and systems with distinctive properties and functions due to their minute scale at the atomic, molecular, or macromolecular levels, typically ranging between 1 and 100 nanometers in length. This scale holds particular significance in heterogeneous catalytic reactions, where active sites comprise tiny crystallites or domains of the active phase (Jin, 2012). The process predominantly involves interactions among atoms and molecules, with active sites located either within or on a large solid substrate. The access paths to these active sites typically have diameters ranging from tens to hundreds of nanometers, making the topic of access paths a prominent area of study in heterogeneous catalysis (Dahman et al., 2022).

Zeolites, commonly utilized as catalysts, owe a significant portion of their unique catalytic capability to their capacity for selectively accommodating molecules based on their shape. This section will utilize specific examples to highlight the impact of nanotechnology on catalyst synthesis and the various methodologies employed in conducting catalytic reactions (Khadzhiev, 2011).

4.1. NANOTECHNOLOGY FOR CATALYSIS

Learning Objectives

- Understand how the properties of catalysts are optimized with the help of nanotechnology.
- Comprehend the advantages of using nanotechnology for creating active metal sites and oxide active phases.
- Understand the significance of nanoreactors in heterogeneous catalysis.

The extensive narrative of the industrialization of catalysis is followed by the advancement of numerous diverse techniques to fabricate catalysts in an endeavor to enhance their qualities. Nevertheless, despite the effort, the achievement of custom designing active sites and site settings with flawless selection and desirable activity remains an unattained objective. Effectively managing a catalytic process is an ongoing and significant difficulty (Yang et al., 2015). The advent of nanotechnology has provided new methodologies that have led to some advancements in the pursuit of this objective. This can be exemplified by instances in the subse□uent three domains:

i. Production of metal active sites and oxides active phases.
ii. Exerting control on the surrounding environment of active sites.
iii. Creating nanoreactors (Banin et al., 2020).

4.1.1. Synthesis of Active Metal Sites and Oxide Active Phases

Commercially relevant sponsored noble metal catalysts typically consist of minute metal crystallites, often measuring less than a few nanometers. The surfaces of these crystallites are enhanced with diverse metal atoms, whose properties vary depending on their positions on the surface, such as corners, edges, or flat areas. These different surface metal atoms exhibit distinct behaviors in reactions, influenced by their structural arrangement. For instance, during the synthesis of ammonia from hydrogen and nitrogen, various surface crystal structure planes of iron metal demonstrate differing reactivity levels (Cuenya, 2010).

Achieving uniform shape and size in metal particles remains a challenging endeavor in the field of metal catalysis. Particularly when the active phase comprises multiple components, it is imperative to produce particles with consistent composition to ensure optimal catalytic performance (An & Somorjai, 2015).

Did You Know?

The active sites in nanocatalysts are often only a few nanometers in size, yet they can drive reactions that convert thousands of molecules per second.

The dendrimer-assisted method presents an innovative approach to synthesizing particles characterized by precise and uniform size distributions. This technique involves a series of intricate steps, outlined in Figure 4.1. Initially, metal ions present in solution are bound to a dendrimer, primarily through the amine groups located on the outer surface of an OH-terminated poly(amidoamine) molecule. Following complexation, these metal ions undergo reduction to individual metal atoms. Subsequently, these atoms amalgamate to form metal particles via agglomeration. The synthesis of dendrimers offers notable advantages owing to their exceptional purity and strong metal ion binding capabilities, rendering them highly efficient in this context (Lee & Yoo, 2014).

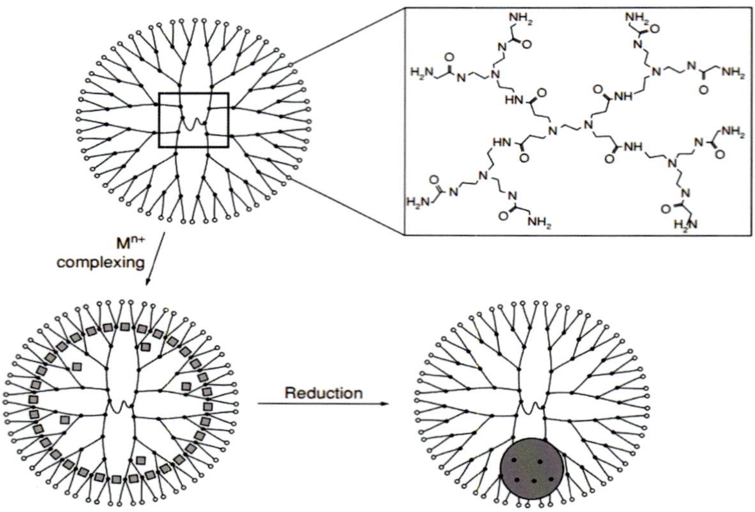

Figure 4.1. *Schematic drawing of dendrimer-assisted preparation of uniform-sized metal particles.*

Source: Harold H. Kung, Creative Commons License.

For example, the utilization of the dendrimer-assisted technique yields platinum (Pt) particles with a notably narrower size distribution compared to those synthesized through conventional methods. By manipulating the quantity of metal binding sites within dendrimers, such as employing different generations of dendrimers, this method facilitates the production of metal particles with consistent sizes across a wide spectrum (Ni & Wang, 2015).

Furthermore, the dendrimer-assisted technique extends to the synthesis of bimetallic clusters. The introduction of the second metal ion can occur via partial displacement of the first metal ion, se□uential addition after the first metal ion, or simultaneous co-complexation with the first metal ion. While these approaches are effective in generating bimetallic particles, ensuring a uniform composition remains a challenge, necessitating the exploration of novel techni□ues or modifications to the existing method (Somorjai et al., 2008).

Additionally, there is a burgeoning interest in fabricating metal oxide particles characterized by homogeneous nano-scale dimensions. If these oxide particles comprise multiple components, the presence of nano-sized particles with distinct components could promote closer contact or improved mixing of the constituents. Conse□uently, the final product may exhibit a more uniform composition and require less rigorous processing conditions, such as lower calcination temperatures (Chavali & Nikolova, 2019).

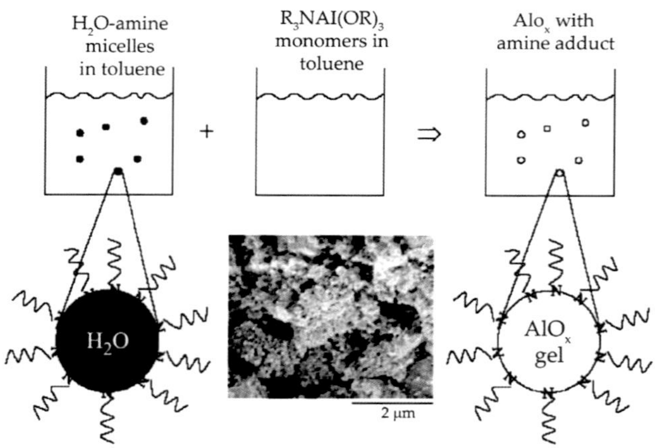

Figure 4.2. *Formation of nanoparticles of metal oxide by reverse by micelle method.*

Source: Mayfair C. Kung, Creative Commons License.

Reverse micelles offer a versatile approach for synthesizing metal oxide nanoparticles. Upon introducing a surfactant molecule with a polar head group and a hydrophobic end into a nonpolar solvent, micelles form, constituting small spherical structures. These micelles consist of a core comprising polar head groups and a shell composed of hydrophobic tails, thus representing reverse micelles (Zaera, 2010). Regulated hydrolysis of the polar solvent, often an alcohol containing an oxide precursor like a metal alkoxide, facilitated by the addition of minimal water, leads to the formation of small particles of oxyhydroxymetal gel within the reverse micelles. Subsequent drying and calcination result in the production of minute metal oxide particles, as depicted in Figure 4.2. The SEM micrograph insert in Figure 4.2 demonstrates aluminum oxide nanoparticles synthesized using this technique. Noteworthy instances, such as minuscule fragments of barium hexaaluminate, have been observed, exhibiting enhanced efficacy in methane combustion compared to conventionally produced counterparts (Makarov et al., 2014).

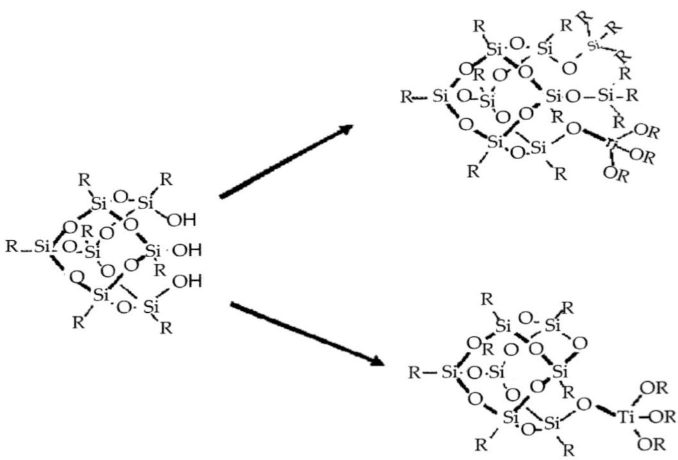

Figure 4.3. *Effect of environment on the catalytic active site.*

Source: Harold H. Kung, Creative Commons License.

In other words, the amine molecules coordinate with the Lewis acid sites of the surface aluminum ions. Surprisingly, this alumina, synthesized and utilized without exposure to temperatures exceeding 70°C, displays a fivefold increase in reactivity in the aminolysis of epoxide compared to traditionally produced sol-gel alumina, which requires activation through heating to 500°C (Mitchell et al., 2021).

4.1.2. Environment Around Active Site

There is a similarity between the effects of binding groups provided by amino acids, which are located far away from the active site of an enzyme, and the well-established concept of steric hindrance, which is generated by ligands and has an effect on the catalytic activity of the metal center in homogeneous catalysis. Recently, there has been a rising understanding of situations in heterogeneous catalysis where the surrounding environment affects an active site in a conceptually similar fashion (Wang et al., 2020).

The structure of modified silses□uioxane, which is represented in Figure 4.3 and shows a partially condensed silsesquioxane, is an example that illustrates this point. Due to the near closeness of the three hydroxyl groups, they can convert into alternative groups by a variety of reactions, including those involving metal alkoxides (Fréchet, 2003). The scenario depicted in Figure 4.3 involves the transformation of one hydroxyl group into a titanium alkoxy group, while the other two hydroxyl groups either undergo a transformation into trimethylsiloxane or unite with each other. Steric hindrance is introduced by the heavy trimethylsiloxane groups, which makes it more difficult for reactant molecules to approach the titanium alkoxide at the same time. As a consequence of this, the catalytic effectiveness of the titanium active center in the process of converting cyclohexene into an epoxide is dramatically reduced, highlighting the critical significance that groups that are not directly linked to the center play (Gentile & de Cortalezzi, 2019).

Did You Know?

Zeolites, which serve as nano reactors in catalysis, are also used in everyday products such as laundry detergents, where they help to soften water and enhance cleaning efficiency.

These examples demonstrate how the environment affects the active sites through the channel of the mesoporous MCM-41 material or the cavity of the molecular sieve. Due to the finite □uantity of these materials, there is a restricted range of conditions that can be utilized or altered (Zhou & Keeling, 2013). Therefore, it is advantageous to have the ability to create cages in a manner that allows for the necessary adaptability to adjust the size of the cage, the size of the access window, and the composition of the cage walls in order to attach functional groups at certain locations. This flexibility would allow for the precise arrangement of various active sites and binding sites for reactants in specified positions relative to each other (Salmeron & Schlögl, 2008). This would permit the activation of specific parts of a reactant molecule. One potential strategy to accomplish this is to construct the cage by adding one siloxane unit at a time. In order to accomplish this, it is necessary to create novel methods for synthesizing the desired compounds. Fortunately, there have been promising advancements made in this area (Mehndiratta et al., 2013). One can now create bicyclic siloxane compounds with the ability to tailor the length of each siloxane bridge. Figure 4.4 depicts a cage known as bicyclo[7.5.3]octasiloxane. This cage consists of three bridges that extend from the bridgeheads. These bridges contain single, double, and triple siloxane units, respectively. A notable feature of this method is its inherent capability to selectively bind functional groups to any of the silicon atoms present on the bridges (Formoso et al., 2016).

4.1.3. Nano Reactors

As the dimension of the cages rises, the cavity reaches a sufficient size to function as a nanoreactor. Nanometer-sized capsules or spheres composed of lipid bilayers have been extensively studied for many purposes, such as drug administration. Nevertheless, unaltered vesicles may not be appropriate as reactors due to the inherent fluidity of the lipid bilayer and the continuous exchange of lipid molecules with free lipids in the surrounding solution. To utilize these vesicles as reactors, it is imperative to transform them into a resilient controller (Centi & Perathoner, 2011).

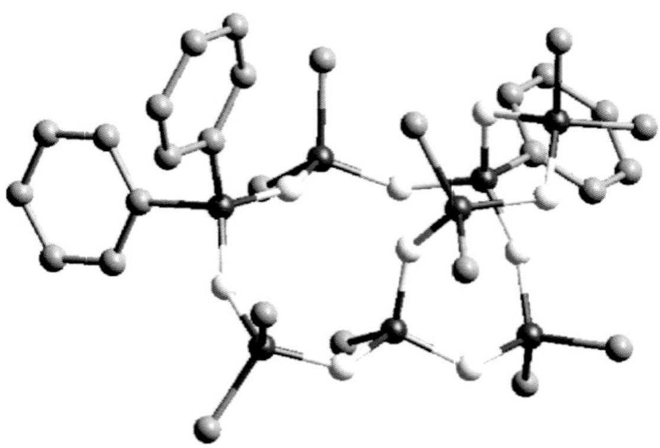

Figure 4.4. *Structure (hydrogen atoms omitted) of a bicycle siloxane,1,5,5-triphenyl-3,3,7,7,9,11,11, 13,13, 16,16-undecamethylbicyclo[7.5.3]octasiloxane, synthesized using a stepwise method..*

Source: Mayfair C. Kung, Creative Commons License.

An effective strategy involves the utilization of lipid molecules containing conjugated carbon double bonds capable of undergoing photopolymerization, such as 10,12-pentacosadiynol. Photopolymerization chemically binds the lipid molecules together, imparting strength and durability to the vesicle structure. Emerging research indicates the feasibility of encapsulating smaller vesicles within larger ones, thereby enabling distinct reactions to occur in separate compartments in a predefined se□uence (Isola et al., 2022).

In addition to mesoporous MCM-41 materials, several methods have been developed for fabricating nano-scale channels and tubes. One such method entails the anodization of alumina with precise control over the experimental conditions. This process yields a thin layer with micrometer-scale dimensions, nanometer-scale diameters, and uniformly designed channels on an aluminum substrate. Subsequently, this layer can be dissolved to produce a thin, porous membrane (Narkiewicz, 2018). The surface of the alumina channel is endowed with hydroxyl groups capable of reacting with metal complexes, leading to the formation of a thin layer of metal oxide. Through iterative implementation of this procedure, precise manipulation of the channel diameter is achieved. The versatility to employ various metal complexes facilitates the creation of channels with diverse oxide coatings, thereby significantly expanding the selection of available oxide channels (Singh & Tandon, 2014).

Carbon nanotubes, possessing nanoscale channels, have attracted considerable attention in recent years, with significant advancements in production methodologies. Currently, researchers are focused on acquiring insights into the separation, control, and mitigation of discrepancies in the characteristics of carbon nanotubes. Their potential for catalytic applications is evident, with one conceivable application being the confinement of catalytic active sites within a carbon nanotube. Furthermore, innovative techniques have emerged to modulate access to nanotube conduits (Tahir et al., 2022). For instance, tube openings can be modified with biotin, capable of binding with streptavidin to restrict the ingress and egress of substances from the channel. The ability to regulate access to a catalytic site, thereby enabling its activation and deactivation, finds successful demonstration in enzyme catalysis. By strategically attaching a temperature-responsive polymer to endoclucanase, it was elucidated that elevating the temperature above the transition temperature results in a significant decrease in enzymatic activity. This phenomenon arises from the transition of the flexible polymer to a crystalline state, obstructing the enzyme's active site and impeding its interaction with reactants. Conversely, cooling the temperature prompts the polymer to revert to its flexible state, thereby restoring enzymatic activity (Kamali et al., 2019).

Practice Problem

Explain how the support material influences the microenvironment around active catalytic sites and affects the performance of the catalyst?

Solution to Practice Problem

The support material plays a crucial role in determining the microenvironment around active catalytic sites. It influences the dispersion, electronic properties, and interaction with reactants.

Key Influences:

Dispersion: A high surface area support material allows for better dispersion of active metal sites, increasing the number of accessible active sites.

Electronic Properties: The electronic properties of the support can affect the electronic state of the active sites, influencing their reactivity.

Interaction with Reactants: The support material can interact with reactants and intermediates, affecting the overall reaction mechanism and efficiency.

4.2. CATALYSIS FOR NANOTECHNOLOGY

Learning Objectives

- Comprehend catalytic processes in the synthesis of nanomaterials, such as nanotubes.

Catalysis has played a pivotal role in the fabrication of carbon nanotubes, addressing hurdles in their integration into electronic applications due to the absence of a dependable production method yielding consistent properties. The distinctive attributes of carbon nanotubes, such as remarkable mechanical strength and customizable electrical conductivity, are influenced by factors like diameter, chirality, and whether they are single-walled or multi-walled (Daniel & Astruc, 2004).

Recent advancements have significantly contributed to surmounting these challenges. Initially, carbon nanotubes were synthesized using methods such as plasma-assisted combustion or hydrogenolysis of carbon-containing compounds (Yan, 2013). However, catalytic hydrogenolysis of hydrocarbons has emerged as a more efficacious approach. The intimate relationship between the diameter of the nanotube and the size of the catalyst particle has facilitated the production of nanotubes with uniform diameters by ensuring consistent generation of catalyst particles. Moreover, manipulating the composition of the catalyst particle enables the fabrication of either single-walled or multi-walled nanotubes (Freund et al., 2014).

SUMMARY

- The recent progress in nanotechnology, especially when combined with a deep comprehension of the methods and phenomena, has created new opportunities for catalytic technology.
- Undoubtedly, there exist numerous prospects for enhancement, particularly in regulating the specifics of the active sites, including their chemical compositions and atomic configurations. However, researchers are finding that the use of inorganic and organic-inorganic hybrid substances provides greater versatility due to their flexibility.
- It should be possible to consider activating and deactivating catalytic active sites as desired, and adjusting the selectivity of a reaction through deliberate design. Although still in its early stages, as nanotechnology expands its range of uses, the potential for influence will increase.
- Nanomachinery is particularly suitable for heterogeneous catalysis due to its ability to harness the energy generated by catalytic events. This domain will likely generate significant attention soon. Environment preparation entails configuring the development environment by installing essential libraries and frameworks like TensorFlow or PyTorch and verifying the presence of a GPU for faster training.

MULTIPLE CHOICE QUESTIONS

1. Which of the following techniques is commonly used for the synthesis of active metal sites in heterogeneous catalysis?

a. Atomic Layer Deposition

b. X-ray Diffraction

c. Scanning Electron Microscopy

d. Ultracentrifugation

2. What role does the environment around an active site play in heterogeneous catalysis?

a. It determines the color of the catalyst.

b. It influences the activity, selectivity, and stability of the catalyst.

c. It controls the weight of the catalyst.

d. It defines the physical state of the catalyst.

3. Which of the following is an example of a nano reactor used in catalysis

a. Zeolites

b. Supercomputers

c. Hydraulic Presses

d. Microscopes

4. How does nanotechnology enhance the field of heterogeneous catalysis?

a. By increasing the weight of the catalysts

b. By providing precision and control in the synthesis of catalytic materials

c. By making the catalysts non-reactive

d. By eliminating the need for catalysts altogether

5. In the context of nano reactors, what is the primary benefit of using confined spaces at the nanoscale?

a. They make the reaction irreversible.

b. They enhance reaction rates and selectivity.

c. They increase the size of the catalyst.

d. They reduce the temperature of the reaction.

REVIEW QUESTIONS

1. What are the main techniques used for the synthesis of active metal sites in heterogeneous catalysis? Discuss the advantages of these techniques?
2. How does the microenvironment around an active site influence the performance of a catalyst?
3. What are the key factors that determine the environment around an active catalytic site?
4. What are the benefits of using confined nanoscale spaces for catalytic reactions?
5. How does catalysis play a role in the synthesis of nanomaterials such as nanoparticles and nanotubes?

Answers to Multiple Questions

1. (a); 2. (b); 3. (a); 4. (b); 5 (b)

REFERENCES

1. An, K., & Somorjai, G. A. (2015). Nanocatalysis I: Synthesis of metal and bimetallic nanoparticles and porous oxides and their catalytic reaction studies. *Catalysis Letters*, *145*, 233–248.
2. Banin, U., Waiskopf, N., Hammarström, L., Boschloo, G., Freitag, M., Johansson, E. M., & Brudvig, G. W. (2020). Nanotechnology for catalysis and solar energy conversion. *Nanotechnology*, *32*(4), 42003.
3. Centi, G., & Perathoner, S. (2011). Creating and mastering nano-objects to design advanced catalytic materials. *Coordination Chemistry Reviews*, *255*(1 3–14), 1480–1498.
4. Chavali, M. S., & Nikolova, M. P. (2019). Metal oxide nanoparticles and their applications in nanotechnology. *SN Applied Sciences*, *1*(6), 607.
5. Cuenya, B. R. (2010). Synthesis and catalytic properties of metal nanoparticles: Size, shape, support, composition, and oxidation state effects. *Thin Solid Films*, *518*(12), 3127–3150.
6. Dahman, Y., Ignatius, N., Poblete, A., Krunic, A., Ma, P., Gosalia, N., & Ali, T. (2022). Nanotechnology: Application and potentials for heterogeneous catalysis. In *Nanotechnology in Paper and Wood Engineering*. Elsevier. (Vol. 2, pp. 231–263).
7. Daniel, M. C., & Astruc, D. (2004). Gold nanoparticles: Assembly, supramolecular chemistry, quantum-size-related properties, and applications toward biology, catalysis, and nanotechnology. *Chemical Reviews*, *104*(1), 293–346.
8. Formoso, P., Muzzalupo, R., Tavano, L., De Filpo, G., & Pasquale Nicoletta, F. (2016). Nanotechnology for the environment and medicine. *Mini Reviews in Medicinal Chemistry*, *16*(8), 668–675.
9. Fréchet, J. M. (2003). Dendrimers and other dendritic macromolecules: From building blocks to functional assemblies in nanoscience and nanotechnology. *Journal of Polymer Science Part A: Polymer Chemistry*, *41*(23), 3713–3725.
10. Freund, H. J., Nilius, N., Risse, T., & Schauermann, S. (2014). A fresh look at an old nano-technology: Catalysis. *Physical Chemistry Chemical Physics*, *16*(18), 8148–8167.
11. Gentile, G. J., & de Cortalezzi, M. M. F. (2019). Nanotechnology and the environment. In *Harnessing Nanoscale Surface Interactions*. Elsevier. (Vol. 2, pp. 41–76).
12. Isola, L. A., Chen, T. C., Elveny, M., Alkaim, A. F., Thangavelu, L., & Kianfar, E. (2022). Application of micro and porous materials as nano-

reactors. *Reviews in Inorganic Chemistry*, *42*(2), 121–136.

13. Jin, R. (2012). The impacts of nanotechnology on catalysis by precious metal nanoparticles. *Nanotechnology Reviews*, *1*(1), 31–56.
14. Kamali, M., Persson, K. M., Costa, M. E., & Capela, I. (2019). Sustainability criteria for assessing nanotechnology applicability in industrial wastewater treatment: Current status and future outlook. *Environment International*, *125*, 261–276.
15. Khadzhiev, S. N. (2011). Nanoheterogeneous catalysis: A new sector of nanotechnologies in chemistry and petroleum chemistry (A review). *Petroleum Chemistry*, *51*, 1–15.
16. Kung, H. H., & Kung, M. C. (2004). Nanotechnology: Applications and potentials for heterogeneous catalysis. *Catalysis Today*, *97*(4), 219–224.
17. Lee, D. W., & Yoo, B. R. (2014). Advanced metal oxide (supported) catalysts: Synthesis and applications. *Journal of Industrial and Engineering Chemistry*, *20*(6), 3947–3959.
18. Makarov, V. V., Love, A. J., Sinitsyna, O. V., Makarova, S. S., Yaminsky, I. V., Taliansky, M. E., & Kalinina, N. O. (2014). "Green" nanotechnologies: Synthesis of metal nanoparticles using plants. *Acta Naturae*, *6*(20), 35–44.
19. Mehndiratta, P., Jain, A., Srivastava, S., & Gupta, N. (2013). Environmental pollution and nanotechnology. *Environment and Pollution*, *2*(2), 49.
20. Mitchell, S., Qin, R., Zheng, N., & Pérez-Ramírez, J. (2021). Nanoscale engineering of catalytic materials for sustainable technologies. *Nature Nanotechnology*, *16*(2), 129–139.
21. Narkiewicz, U. (2018). Nano-catalysts: Research–technology–industrial applications. *Nanoscience and Nanotechnology*, *4*(3), 165–170.
22. Ni, B., & Wang, X. (2015). Face the edges: Catalytic active sites of nanomaterials. *Advanced Science*, *2*(7), 1500085.
23. Salmeron, M., & Schlögl, R. (2008). Ambient pressure photoelectron spectroscopy: A new tool for surface science and nanotechnology. *Surface Science Reports*, *63*(4), 169–199.
24. Singh, S. B., & Tandon, P. K. (2014). Catalysis: A brief review on nano-catalyst. *J Energy Chem Eng*, *2*(3), 106–115.
25. Somorjai, G. A., Tao, F., & Park, J. Y. (2008). The nanoscience revolution: Merging of colloid science, catalysis and nanoelectronics. *Topics in Catalysis*, *47*, 1–14.
26. Tahir, M. B., Sohaib, M., Sagir, M., & Rafi□ue, M. (2022). Role of nanotechnology in photocatalysis. *Encyclopedia of Smart Materials*, *7*(2), 578–580.

27. Wang, X., Jia, Y., Mao, X., Liu, D., He, W., Li, J., & Yao, X. (2020). Edge-rich Fe– N4 active sites in defective carbon for oxygen reduction catalysis. *Advanced Materials*, *32*(16), 2000966.
28. Yan, N. (2013). When nanotechnology meets catalysis. *Nanotechnology Reviews*, *2*(5), 485–486.
29. Yang, F., Deng, D., Pan, X., Fu, Q., & Bao, X. (2015). Understanding nano effects in catalysis. *National Science Review*, *2*(2), 183–201.
30. Zaera, F. (2010). The new materials science of catalysis: Toward controlling selectivity by designing the structure of the active site. *The Journal of Physical Chemistry Letters*, *1*(3), 621–627.
31. Zhou, C. H., & Keeling, J. (2013). Fundamental and applied research on clay minerals: From climate and environment to nanotechnology. *Applied Clay Science*, *74*, 3–9.

CHAPTER 5

NANOTUBES FOR ENERGY STORAGE

LEARNING OBJECTIVES

At the end of this chapter, readers will be able to:

1. Understand the mechanism of hydrogen storage
2. Comprehend different nanotubes for hydrogen storage
3. Understand how CNTs improve the performance of lithium-ion batteries as anode materials
4. Comprehend the role of titanium dioxide (TiO) nanotubes in lithium-ion batteries.

INTRODUCTORY EXAMPLE

Arkema, a global specialty chemicals company, has been actively involved in the research and development of advanced materials, including nanotubes, for energy storage applications. Their work highlights the practical uses of carbon nanotubes (CNTs) and inorganic nanotubes in both hydrogen storage and lithium batteries.

Arkema has developed carbon nanotube-based materials for efficient hydrogen storage. The unique properties of CNTs, such as their high surface area and ability to adsorb hydrogen at room temperature, make them ideal for this application. Arkema's research focuses on optimizing the synthesis and functionalization of CNTs to enhance their hydrogen storage capacity. These advancements could lead to more practical and efficient hydrogen storage systems for use in fuel cell vehicles and other renewable energy applications.

Arkema has also been exploring the use of CNTs in lithium-ion batteries. Their pure CNT anode materials offer excellent electrical conductivity and mechanical strength, which improve the overall performance and lifespan of the batteries. Arkema's CNT-based anodes provide higher energy density and faster charge-discharge cycles compared to traditional graphite anodes, making them suitable for a range of applications from consumer electronics to electric vehicles.

Arkema's advancements in nanotube technology for energy storage demonstrate the significant potential of these materials in addressing modern energy challenges.

UNIT INTRODUCTION

Energy has played a crucial role in the advancement of every aspect of human society. The global energy requirements, including power, transportation, and heating fuels, are projected to increase twofold by 2050 and more than triple by the close of the century. This surge in demand raises significant concerns over the availability of energy supply. Efficiently overseeing the utilization, conveyance, and preservation of energy is a necessity in any society. In the long run, conventional energy sources like fossil fuels and coal will not be sufficient to meet the demand (Yan et al., 2015). Moreover, their ongoing utilization leads to detrimental consequences such as pollution, which poses a threat to human well-being, and the emission of greenhouse gasses that contribute to climate change. Adequate quantities of renewable energy are closely connected to worldwide stability, economic affluence, and quality of life. Discovering energy sources to meet the increasing global demand is one of the most significant challenges facing society in the next 50 years (Dillon, 2010). Solar energy and hydrogen are seen as prominent contenders because of their abundant, clean, and renewable nature. Various technologies have been created to harness solar energy, including solar electricity (photovoltaic cell, PV), solar fuels (photoelectrochemical cell, PEC), and solar thermal systems. The photovoltaic (PV) cell transforms solar energy into electricity, while the photoelectrochemical (PEC) process separates water into hydrogen and oxygen, so converting solar energy into chemical energy (Samy et al., 2020).

Thermal energy is derived from the sun in thermal systems. Without thorough examination and exploration of energy storage technologies, the existing clean energy technologies would face growing challenges in delivering dependable and cost-effective electricity. This is due to the need to dispose of and squander extra energies, such as electricity and hydrogen. Enhanced energy storage technology would significantly increase the utilization of transient renewable energy sources (Wen et al., 2016).

Energy storage technologies offer a solution to the increasing constraints of energy supply and demand and improve the dependability of the electric grid. Presently, there exist two types of large-scale energy storage systems: electrochemically based systems, such as advanced lead-carbon batteries, lithium-ion batteries, sodium-based batteries, flow batteries, and electrochemical capacitors; and kinetic-energy-based systems, such as compressed-air energy storage and high-speed flywheels (Shi et al., 2014). Presently, the existing energy storage technologies fail to sufficiently fulfill the diverse re□uirements of the electric power business because of the high cost, limited energy capacity, low efficiency, and inherent complexity of most contemporary storage systems. To overcome the challenges related to the economic and technical aspects of energy storage systems, as well as to fully utilize the potential of large-scale energy storage technologies, it is crucial to make significant progress in the development of materials and devices. This is because the performance specifications of these technologies are largely determined by the materials they are made of. The choice of materials should ensure that the storage solutions are cost-effective, efficient, dependable, and environmentally friendly. There is now a widespread recognition that advancements in technology and groundbreaking innovations are necessary (Ma et al., 2008).

Nanoscience and nanotechnology emerge from the utilization of unique characteristics, phenomena, procedures, and capabilities that matter demonstrates at sizes ranging from isolated atoms or molecules (about 1 nm) to larger bulk materials (over 100 nm). The incorporation of nanoscience and nanotechnology is essential for the advancement of energy storage technologies, as all fundamental processes involved in energy storage, including charge transfer, molecular rearrangement, and chemical reactions, occur at the nanoscale (Wang et al., 2003). The emergence of novel nanoscale materials and structures presents a fresh framework for the advancement of innovative energy storage technologies such as enhanced batteries, fuel cells, supercapacitors, and biofuels. Nanotubes, also known as NTs, are a novel type of material that has had a significant influence on various applications ever since Iijima discovered the production of carbon nanotubes (CNTs) in 1991. Nanotubes have proven to be advantageous in the field of energy storage because of their distinctive qualities, including their one-dimensional cylindrical tubule structure, large surface area, excellent conductivity, low density, high stiffness, and strong tensile strength. This article aims to provide a comprehensive overview of the latest advancements in energy storage using nanotubes (Zhang et al., 2011).

5.1. HYDROGEN STORAGE

Learning Objectives

- Understand the primary mechanisms by which carbon nanotubes (CNTs) store hydrogen
- Comprehend the structural characteristics of CNTs that influence their hydrogen storage capacity
- Understand the potential of CNT/oxide composites in enhancing hydrogen storage efficiency

Hydrogen is an excellent candidate for a clean energy carrier in the future. It can be generated from various energy sources, possesses the maximum energy density per unit mass, and causes minimal pollution. This is because hydrogen can be taken from natural resources like water or biomass, and its utilization results in the production of water alone. Shifting from fossil fuels to hydrogen as a primary fuel source during the next 50 years could enhance energy security. The extremely low density of hydrogen is a significant obstacle in the creation of a hydrogen storage system that is cost-effective, dependable, and safe (Brzhezinskaya et al., 2019).

For H_2-O_2 fuel-cell driven vehicles to travel 300 miles without refueling, a storage system with a hydrogen density of at least 49 wt% (81 gL^{-1}) is necessary, taking into account the materials and tank used. Additional criteria consist of the ability to be reversed for 1500 cycles, a cost of $2 per kilowatt-hour, a fueling time of 30 seconds per kilogram of hydrogen, and an operating temperature/pressure range of -40 to 60°C and less than 100 atmospheres. The prevailing techniques for hydrogen storage are compressed gas and liquid (Rashidi et al., 2010). Prototype hydrogen-powered vehicles are already utilizing carbon fiber-reinforced compressed hydrogen gas tanks with pressure ratings of 5,000-psi and 10,000-psi. Nevertheless, the capacity of hydrogen they can contain is insufficient, being below 4.5 wt% hydrogen (36 g L^{-1}). Storing hydrogen in a cryo-compressed tank can increase its volumetric density by two times. At a temperature of 21 K, a weight percentage of 5 wt% (78 g L^{-1}) has been successfully attained. Nevertheless, the process of converting hydrogen into a liquid state would utilize approximately 30% of its energy content, and the ongoing evaporation of hydrogen would also diminish its total efficiency. The ultimate objective is to create reversible solid-state hydrogen storage materials, specifically high surface area adsorbents (Chen et al., 2004).

Storing hydrogen poses a complex and thrilling problem. Discovering the secrets of storing hydrogen in nanoscale materials could perhaps pave the way for a hydrogen-powered future and help achieve the realistic goals set for 2025 (Figure 5.1). An ideal hydrogen storage material should possess the following characteristics: a) a large capacity in terms of volume or weight, b) a rapid sorption rate at relatively low temperatures, and c) a strong resistance to degradation during recycling (Froudakis, 2011). The primary approach in storing H_2 in solid sorbents involves several strategies: (1) increasing the specific surface area, (2) modifying the pore structure to enhance the binding energy of hydrogen molecules, with defects and appropriate pore structure leading to a 20–30% increase in binding energy compared to perfect or large pore structures, (3) introducing more electronegative atoms like nitrogen (N) to strengthen the bonding energy, and (4) depositing nanosized transition metal catalysts to facilitate chemisorption through reversible hydrogen spillover. In this method, a catalyst breaks down molecular hydrogen into hydrogen atoms, which then diffuse into the sorbent and form reversible bonds with carbon atoms or other atoms in the sorbent (Gordon & Saeger, 1999).

Did You Know?

The high surface area of carbon nanotubes allows for significant hydrogen adsorption, which is crucial for efficient hydrogen storage systems.

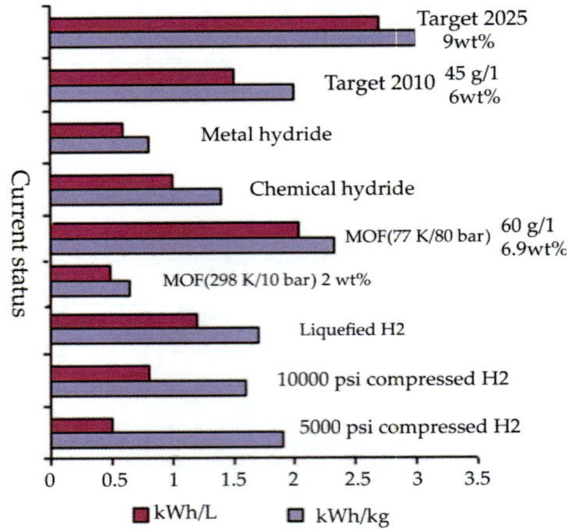

Figure 5.1. *Representation of the current status of today's hydrogen storage technologies in volumetric and gravimetric terms.*

Source: Gang Zang, Creative Commons License.

5.1.1. Carbon Nanotubes

Carbon nanotubes (CNTs) are cylindrical structures formed by rolling up two-dimensional graphene sheets. Carbon nanotubes (CNTs) can exist in three different forms based on the number of walls they possess: single-walled (SWCNT), double-walled (DWCNT), and multi-walled (MWCNT). Single-walled carbon nanotubes (SWCNTs) typically have a diameter of 1–2 nanometers and can reach lengths of up to several microns. They are frequently combined because of the van der Waals forces of attraction. Multi-walled carbon nanotubes (MWCNTs) typically possess a hollow center measuring tens of nanometers in diameter, with a gap between adjacent layers of 0.34–0.37 nm. The inner tube diameter of a double-walled carbon nanotube (DWCNT) falls between the range of 1–3 nm, while its interlayer spacing is within the range of 0.3–1 nm (Cao & Wei, 2013).

5.1.1.1. Functionalized/Doped Carbon Nanotube

Functionalization is a deliberate process that can impart impurities or functional groups to carbon nanotubes (CNTs) to increase their capacity to absorb hydrogen. Nevertheless, the presence of impurities in carbon nanotubes (CNTs) has been identified as a contributing factor in the variation in hydrogen absorption capacity observed in CNTs, as described in the literature. For instance, the K-modified single-walled carbon nanotubes (SWCNT) exhibit an adsorption capacity of less than 0.69 weight percent (wt%) at room temperature and 70 bar pressure, and less than 1.05 wt% at 77 K temperature and 50 bar pressure. Hirscher et al. and Costa et al. have demonstrated that the presence of metal alloy particles in the single-walled carbon nanotubes (SWCNTs) is the only factor responsible for the absorption of hydrogen by the decorated SWCNTs (Tuci et al., 2014). The reversible hydriding/dehydriding of these alloys explains the majority of the detected desorption volumes in this investigation. However, transition metals are frequently documented as catalysts that play a crucial role in enhancing hydrogen storage and improving kinetics. Yildirim et al. showed through first-principle calculations that a single titanium (Ti) atom coated on a single-walled carbon nanotube (SWCNT) can attach to as many as four hydrogen molecules. They also predicted that SWCNTs with high coverage of Ti can adsorb up to 8 weight percent (wt%) of hydrogen. H_2 is dissolved by metallic Pd in Pd-decorated MWCNTs, resulting in the formation of H atoms or protons. These particles can quickly permeate the walls of the tubes and unite to form H_2 within the interlayers and hollow interiors. The carbon nanotubes (CNTs) that were adorned with palladium (Pd) demonstrated a hydrogen storage capacity of 3.5 weight percent (wt%) under ambient temperature and a pressure of 100 bar. This value is almost seven times greater than the hydrogen storage capacity (0.6 wt%) of the CNT samples that were not adorned with palladium. In general, the absorption of hydrogen in carbon nanotubes (CNTs) is increased when they are adorned with metal. The elevated desorption activation energy contributes to an increase in the release temperature (Zikalala et al., 2018).

A proposal has been made to elucidate and control the process of hydrogen chemisorption on carbon nanotubes (CNTs) in the presence of a metal catalyst. This proposal involves the concept of reversible hydrogen spillover. The catalyst within the nanotube structure causes the separation of molecular hydrogen. Due to the curvature of the nanotube, the attractive forces between the carbon atoms in the lattice coincide. Consequently, there is an increased number of carbon atoms engaging with the hydrogen, leading to a reduction in the energy required for hydrogen addition and enabling the hydrogen to move away from the catalyst through surface diffusion. Atomic hydrogen migrates from the catalyst to the single-walled carbon nanotube (SWCNT) where it is retained (Suryanto et al., 2016). If the binding energy between carbon and hydrogen is relatively small, the process may be reversible. Figure 5.2 illustrates the dissociative adsorption of H_2 molecules on the surface of the Pt catalyst. Subsequently, atomic hydrogen atoms migrate to the active carbon material, which serves as the substrate for the Pt catalyst. The physical connections between the active carbon and SWCNTs facilitate the transfer of H atoms from the Pt/active-carbon catalyst to SWCNTs. When it comes to single-walled carbon nanotubes (SWCNTs), hydrogen atoms attach to different areas than hydrogen molecules (H_2). The strength of the bond between atomic hydrogen and SWCNT is quite high, which is supported by the observed hysteresis (Xiang et al., 2016).

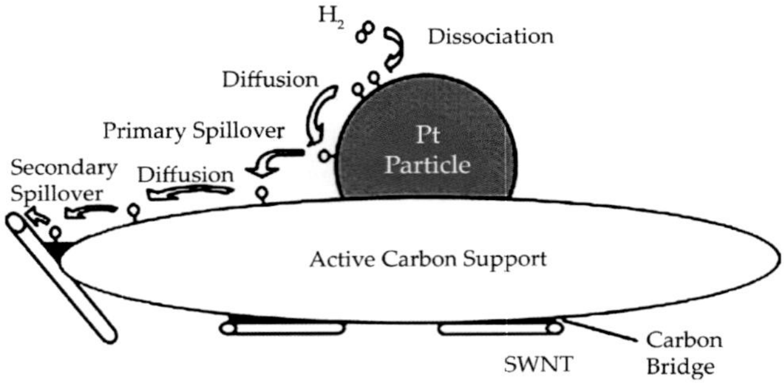

Figure 5.2. Schematic illustration of the spillover mechanism.

Source: Navin Manjooran, Creative Commons License.

The subsequent analysis of previously adsorbed samples indicated that around 70–85% of the hydrogen that was spilled fills the physisorption binding sites. The hydrogen that is left (30–15%) could potentially chemisorb onto faulty sites and cannot be removed reversibly (Uceta et al., 2023).

5.1.2. Inorganic (Non-Carbon) Nanotubes

A proposal was made that suggests that nearly every sheet-like material can spontaneously form scrolls through the process of exfoliating bulk stacked materials into individual sheets. If the two-dimensional (2D) sheet is large enough, the energy obtained from non-covalent binding between layers of the scroll is enough to offset the energy required to build the scroll. Several materials with several layers have shown their capacity to create structures like nanotubes or layers resembling onions. The materials mentioned are hexagonal BN, GaN, transition metal dichalcogenides such as TiS_2, WS_2, and MoS_2, metal halides like $NiCl_2$, $MgCl_2$, and $CdCl_2$, and metal oxides including MnO_2, V_2O_5, and TiO_2. Inorganic nanotubes can store hydrogen in various locations, including between layers of the sheet, on the exterior surface of the scroll, at the center of the scroll, and in the interstitial space between scrolls. These nanotubes are highly regarded as materials with great potential for hydrogen storage (Martincic et al., 2019).

5.1.2.1. Silicon Carbide Nanotube

The investigation on BNNTs for hydrogen adsorption demonstrated that the presence of point charges on the surface of the material can enhance the storage capacity by increasing the binding energy. Theoretical investigations on silicon carbide nanotubes (SiCNTs) have revealed that the inter-wall spaces of SiCNTs contain point charges due to the transfer of more than half an electron from silicon (Si) to carbon (C). These nanotubes, which have point charges on their surface, are excellent materials for hydrogen storage. A computational analysis using ab initio density functional theory (DFT) revealed a 20% enhancement in the binding energy of hydrogen (H) in SiC nanotubes compared to pure carbon nanotubes (CNTs). This phenomenon can be elucidated by the presence of different charges within the walls of the SiC nanotube (Itas et al., 2023). A conventional Monte Carlo simulation of SiC nanotube bundles showed a significant enhancement in the storage capacity of SiC nanotubes, particularly under low temperature and high-pressure circumstances. The results suggest that SiC nanotubes, similar to other inorganic nanostructures, may be more effective for hydrogen storage compared to pure carbon nanotubes. Li et al. demonstrated through DFT simulations that Li-doping can enhance the binding energy. The adsorption energy of hydrogen molecules on Li-doped SiCNT is 0.211 eV, which is more than the adsorption energy on pure SiCNT (0.086 eV). This difference is attributed to the transfer of charge from Li to the nanotube. The researchers discovered that a maximum of four molecules can be connected to a lithium-adsorbed silicon carbon nanotube (SiCNT), with an average binding energy of 0.165 electron volts (eV) (Mahdizadeh & Goharshadi, 2014).

Several experimental approaches have been devised to produce SiC nanostructures. While there is no available experimental evidence on hydrogen storage on SiCNTs, Pol et al. have reported that large surface area SiC nanorods can store 2.5 wt% H_2 at ambient temperature and 60 bar. Like other nanostructured compound materials, the adsorption and desorption curves in the pressure-composition-isotherm do not align when the pressure is reduced to the ambient level. Only hydrogen gas with a concentration of 1.5 weight percent can undergo reversible absorption and release. These findings indicate that there are chemisorptions of H_2 on SiC nanorods, which could explain why the adsorption at normal temperature is irreversible (Okoroanyanwu et al., 2021).

Did You Know?

Using carbon nanotubes in energy storage applications can contribute to reducing greenhouse gas emissions by improving the efficiency of renewable energy systems and electric vehicles.

5.1.2.2. Tungsten Carbide Nanotube

Tungsten carbide has garnered scientific attention due to its exceptional features, including a high melting point, outstanding hardness, low friction coefficient, great oxidation resistance, and excellent electrical conductivity. Tungsten carbide nanotubes (WCNTs) can be produced through the thermal breakdown of $W(CO)_6$ in the presence of Mg powder. Pan et al. concluded that WCNTs are potentially more appropriate for hydrogen storage due to their somewhat elevated, yet not excessively high, binding energy as determined using first-principles calculations. The WCNT, which is composed of alternating carbon (C) and tungsten (W) atoms, has the highest level of stability according to Figure 5.3. After examining all potential adsorption sites and orientations of H_2 molecules, Pan et al. discovered that the majority of sites are not stable for H_2 adsorption, except the W top site when H_2 is parallel to the tube (Figure 5.4) (Fan et al., 2015). The bond length of H_2 increased to 0.875 Å, which is more than the bond length in an isolated H_2 molecule (0.754 Å). The length of the W–H bond is 1.85 Å. The adsorption of a hydrogen molecule on a zigzag single-wall carbon nanotube (10, 0) was investigated using the same method for comparison. The most stable location for a hydrogen molecule is at the core of the carbon hexagon. The distance of binding and the energy associated with it are 3.4 Å and 0.01 eV, respectively (Figure 5.4). The H_2 binding distance on the WCNT is significantly shorter compared to that on CNT, BNNT, and SiCNT, measuring approximately 3.0 Å. The binding energy of H_2 on the WCNT is approximately 0.44 eV, significantly higher than that on CNT (0.06 eV), BNNT (0.09 eV), and SiCNT (0.07 eV) (Zhao et al., 2021).

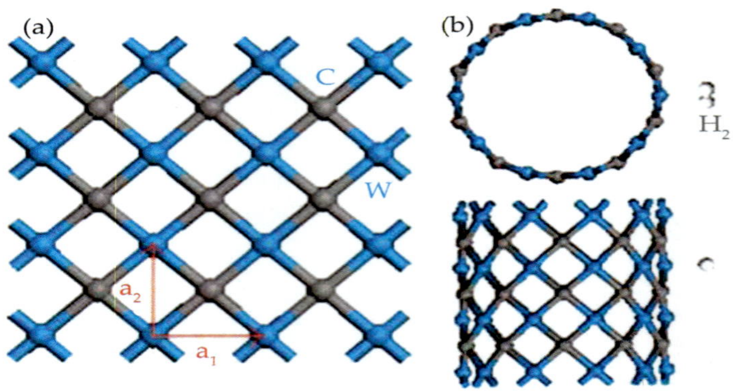

Figure 5.3. *Atomic configuration of (a) tungsten carbide sheet; and (b) a tungsten carbide nanotube with a hydrogen molecule.*

Source: Hui Pan, Creative Commons License.

The shorter binding distance and higher binding energy indicate that the hydrogen storage capability of WCNT is better than that of other nanotubes (Soares et al., 2019).

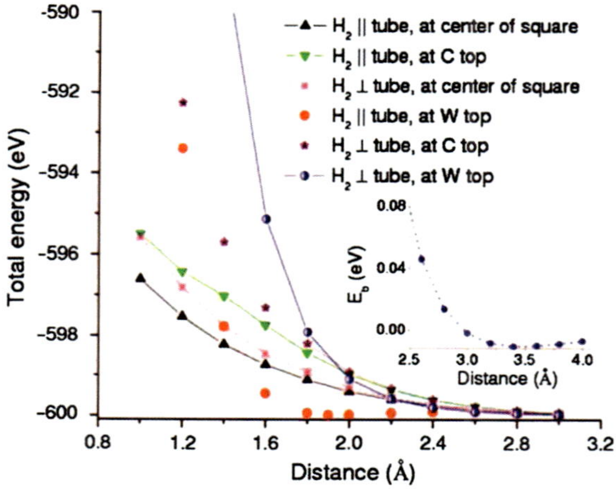

Figure 5.4. *Total energy of a hydrogen molecule as a function of its distance from the outer wall of the WCNT (10, 0).*

Source: Gang Zang, Creative Commons License.

5.1.2.3. Graphitic Carbon Nitride Nanotube

Carbon nitride can exist in multiple phases, which are determined not only by the ratio of carbon to nitrogen but also by the arrangement of atoms. Out of these stages, a type of graphitic carbon nitride called heptazine-based g-C_3N_4 (shown in Figure 5.10(a)) is considered the most stable structure. It may be created by heating and condensing precursors that contain carbon, nitrogen, and hydrogen. A unique nanotube, known as g-C_3N_4 nanotube, was suggested as a photocatalyst for the process of water splitting (Lin et al., 2020). This nanotube is created by curling up a single layer of g-C_3N_4, as shown in Figure 5.5(b). The nanotube's distinct porous structure, including a pore diameter of 0.7 nm (as shown in Figure 5.5), enables enhanced hydrogen penetration into the interior of the nanotube. This, in turn, increases the specific surface area of the nanotube, facilitating hydrogen storage within and thus improving the storage capacity. The N atoms that are bound together in pairs at the borders of the pore serve as active sites for the chemisorption of H and the functionalization of metals (Koh et al., 2012).

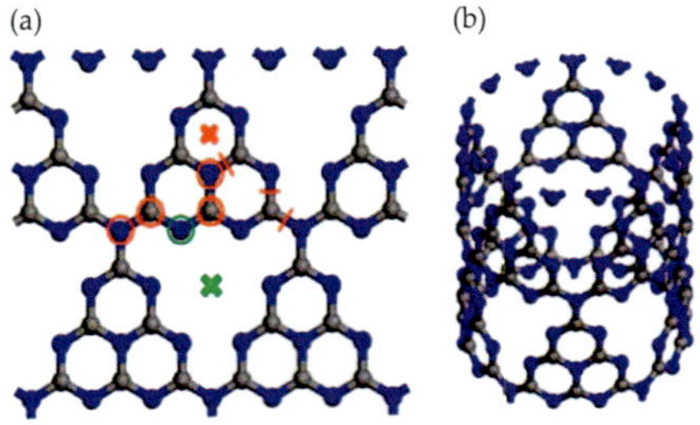

Figure 5.5. *Structure of a heptazine-based graphitic carbon nitride (a) monolayer; and (b) nanotube. C and N atoms are represented in gray and blue respectively.*

Source: Elsevier, Creative Commons License.

5.1.3. CNT/Oxide Composite

Carbon nanotubes (CNTs) can store hydrogen molecules through physical adsorption, making them suitable as reversible hydrogen storage materials. However, the capacity is restricted to a tiny quantity (1 wt%) within the range of ambient temperature and high pressure.

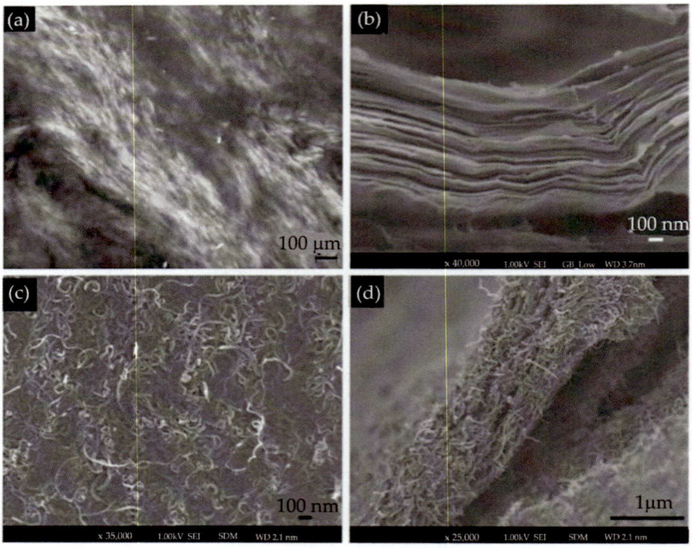

Figure 5.6. *Illustration of a) polarized optical micrograph of LC graphene oxide dispersion and SEM micrographs of; b) GO paper; c) MWCNTs distributed on GO layer; and d) layer-by-layer assembled GO platelets decorated by MWCNTs.*

Source: Wiley, Creative Commons License.

Metal, oxide, and compounds can store hydrogen by undergoing chemical reactions and forming bonds, resulting in a potentially large storage capacity that is achieved at high release temperatures. Nanostructured composite materials consisting of carbon nanotubes (CNTs) and metal or metal oxide nanoparticles or oxides are of great interest for hydrogen storage. This is due to the non-classical s-p-d hybridization and spillover processes, which have the potential to increase the capacity for storing hydrogen (Jampani et al., 2013).

The hydrogen capacity of GO-MWCNTs exceeds the combined capacity of its component constituents, namely GO and MWCNTs, reaching 2.6 wt% (Figure 5.7) (Azam et al., 2013).

Metal-organic frameworks (MOFs) are a swiftly expanding category of porous crystalline solids formed by linking metal ions or clusters with organic molecule bridges. Up till now, all MOFs that have been studied have been found to have microporous architectures, meaning that their cavities are smaller than 2 nm in size. Metal-organic frameworks (MOFs) exhibit a substantial surface area ranging from 1,000 to 5,900 m^2g^{-1} and contain a significant □uantity of accessible sites for the adsorption of atoms or molecules. MOFs, in comparison to commonly used microporous inorganic materials like zeolites, offer greater flexibility in structural design by allowing control over the architecture and functionalization of the pores.

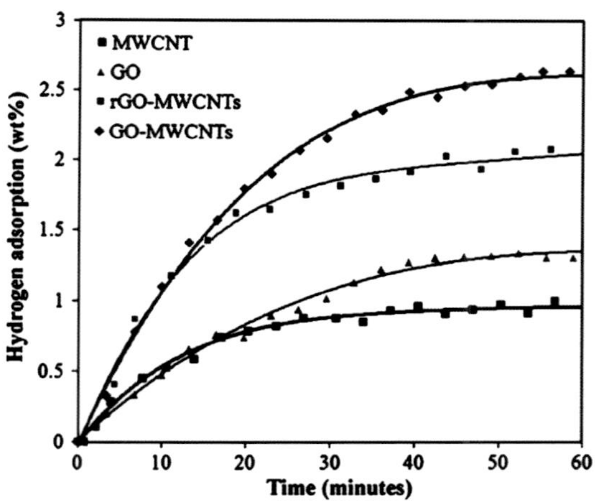

Figure 5.7. *Hydrogen adsorption of GO, MWCNTs, GO-MWCNTs and rGOMW-CNTs as a function of time.14.*

Source: Wiley, Creative Commons License.

The capacity of H_2 storage can be significantly increased by the ability of H_2 to bind to metal atoms in MOFs, compared to simple H_2 adsorption at nonmetal sites in previously produced materials. Nevertheless, the storage of H_2 by MOFs at ambient temperature is significantly lower compared to that seen at 77 K and 1 bar, which is equivalent to 2.5 wt% (Arya et al., 2023).

The resulting composite has an identical crystal structure and morphology to the original MOF-5. However, it showed a significantly higher Langmuir-specific surface area, increasing from 2,160 to 3,550 m^2g^{-1}. Additionally, there was a 50% increase in hydrogen storage capacity, with levels rising from 1.2 to 1.52 wt% at 77 K and 1 bar, and from 0.3 to 0.61 wt% at 298 K and 95 bar. Furthermore, the composite displayed enhanced stability when exposed to ambient moisture (Kapoor et al., 2023).

Practice Problem

Compare the advantages and disadvantages of using inorganic nanotubes, such as boron nitride nanotubes, over carbon nanotubes for hydrogen storage applications.

Solution to Practice Problem

Inorganic nanotubes, such as boron nitride nanotubes, offer several advantages for hydrogen storage, including higher thermal and chemical stability compared to carbon nanotubes. They also have unique adsorption properties that can potentially enhance hydrogen storage capacity. However, disadvantages include more complex and expensive synthesis processes, which can limit their practical applications. Additionally, the storage capacities of inorganic nanotubes may still be lower than the optimized carbon nanotubes, and further research is needed to fully exploit their potential.

5.2. LITHIUM BATTERY

Learning Objectives

- Understand the role of CNTs in Lithium-ion batteries
- Comprehend the role of titanium dioxide (TiO_2) nanotubes in lithium-ion batteries, including their contribution to cycling stability and rate capability.
- Assess the challenges and strategies involved in using silicon nanotubes as anode materials in lithium-ion batteries

A lithium-ion battery (LIB) is a type of rechargeable battery that operates by the movement of lithium ions from the anode to the cathode during discharge and in the opposite direction during charging. Li-ion batteries utilize an intercalated lithium compound as the electrode material, in contrast to the metallic lithium employed in a non-rechargeable lithium battery. Lithium-ion batteries (LIBs) consist of four essential components: anode, cathode, separator, and electrolyte (Figure 5.8). The cathode, usually made of lithium metal oxide, serves as the positive electrode of the battery during discharge, while the anode, often constructed of graphitic carbon, serves as the negative electrode. Li^+ ions migrate between the anode and cathode through the electrolyte during the process of charging and discharging (Shrivastav et al., 2019).

$$\text{Cathode}: \mathrm{LiMO}_n = \mathrm{Li}_{1-x}\mathrm{MO}_n + x\mathrm{Li}^+ + xe^-$$

$$\text{Anode}: x\mathrm{Li}^+ + xe^- + 6\mathrm{C} \leftrightarrow \mathrm{Li}_x\mathrm{C}_6$$

The electrolyte commonly consists of a lithium salt, such as $LiPF_6$, dissolved in an organic solvent, such as ethylene carbonate. This configuration hinders the movement of free electrons between the electrodes. Therefore, the electrons undergo the half-reaction by traversing through an external wire, resulting in the flow of electric current (Tang et al., 2013). Although graphite-based lithium-ion batteries (LIBs) exhibit excellent performance, there are numerous obstacles in the design of LIBs, particularly in enhancing the capacity and cycle life of the anode materials. Commercial lithium-ion batteries (LIBs) often use graphite as the material for the anodes. This is because graphite has excellent electronic conductivity in the plane, due to its delocalized p-bonds. Additionally, graphite has little expansion when lithium is inserted, and its structure is suitable for lithium-ion intercalation and diffusion. When lithium is inserted into graphite, it takes up a position between two layers of graphite, blocking other lithium ions from occupying nearby positions. This restriction results in a maximum of one lithium ion for every six carbon atoms, known as LiC_6. As a consequence, the theoretical specific capacity is limited to 372 mAh g^{-1}, while the observed capacity ranges from 280 to 330 mAh g^{-1}, depending on the type of graphite used (Masih-Tehrani et al., 2013).

Given that the performance of lithium-ion batteries (LIBs) is heavily influenced by the characteristics of the electrodes, it is crucial to make substantial enhancements in the electrochemical properties of electrode materials to achieve the high standards of these applications. This is particularly important because graphite, which is commonly used as an electrode material, has a relatively limited capacity for lithium insertion (Kurzweil, 2015). The hollow nanotubes are highly suitable for use as electrodes in LIBs due to several reasons. Firstly, their large surface area enhances the contact between the electrode and electrolyte. Secondly, their reduced size reduces the distance that Li^+ ions need to travel.

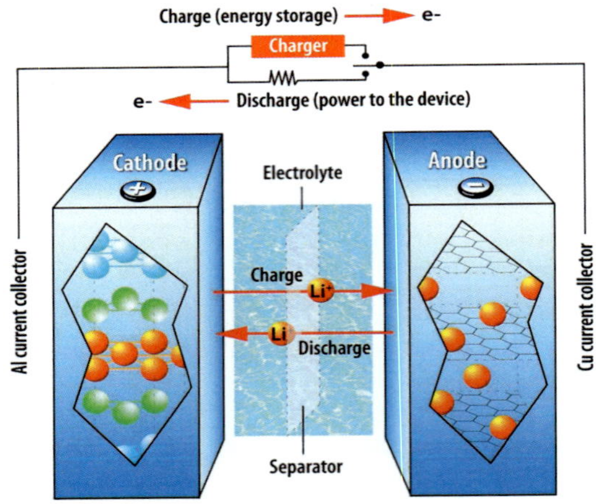

Figure 5.8. *Schematic of lithium-ion cell.*

Source: Navin Manjooran, Creative Commons License.

Thirdly, the space within the hollow structures helps to prolong the cycling life of the electrodes by mitigating the volume change that occurs during Li insertion and desertion, thereby preventing pulverization and aggregation of the electrode materials. Lastly, the additional space provided by the hollow structure allows for increased storage of Li^+ ions, leading to enhanced specific capacity (Thackeray et al., 2012).

5.2.1. Pure CNTs

The distinctive characteristics of carbon nanotubes (CNTs), including their hollow structure, excellent conductivity, low density, high stiffness, and strong tensile strength, make them highly suitable for use as anode electrodes in lithium-ion batteries (LIBs). The structural and morphological properties of carbon nanotube (CNT) based anode electrodes have a significant impact on their performance in lithium-ion batteries (LIBs), similar to their effects on hydrogen storage and supercapacitors. These properties include the chirality of single-walled CNTs (SWCNTs), the type of CNTs (SWCNTs or multi-walled CNTs), defects, diameter, length, and alignment. The anode materials in LIBs must possess high electrical conductivity to facilitate the rapid movement of charge carriers from the current collector to the electrolyte (Ge et al., 2014). The electrical properties of single-walled carbon nanotubes (SWCNTs) can vary depending on their chirality, with some exhibiting semiconducting behavior while others display metallic behavior. Kawasaki et al. observed that the metallic single-walled carbon nanotubes (SWCNTs) have a Li-ion storage capacity that is approximately five times higher than that of the semiconducting nanotubes. Chew et al. found that the electrochemical performance of multi-walled carbon nanotubes (MWCNTs) is superior to that of single-walled carbon nanotubes (SWCNTs) (Bai et al., 2019). Additionally, the flexible MWCNT electrodes exhibit consistent cycling behavior and can handle a charge-discharge rate of up to 10C. Eom et al. found that the defective multi-walled carbon nanotubes (MWCNTs) had a reversible capacity of 681 $mAhg^{-1}$, which is significantly larger than the pristine MWCNTs with a capacity of 351 $mAhg^{-1}$. This is due to the presence of defects in the nanotubes, which promote the insertion of lithium ions into the tube and improve the overall capacity for reversible lithium storage (Rakhi et al., 2012).

Remember

Titanium dioxide nanotubes are valued for their cycling stability and rate capability, making them suitable for use in lithium-ion batteries despite their lower theoretical capacity.

The CNT-based anodes have significantly non-reversible capacity in comparison to graphite-based anode materials due to their large surface area, hollow structure, and smaller size. The electrochemical performance of carbon nanotube (CNT) based anodes in lithium-ion batteries (LIBs) is significantly influenced by their shape, structure, and alignment, as well as the synthesis and treatment procedures used during manufacture. To enhance their performance as anodes, it is necessary to optimize the diameter, introduce flaws, reduce the length, align the carbon nanotubes (CNTs), and improve the conductivity (Sun et al., 2017) (Figure 5.9).

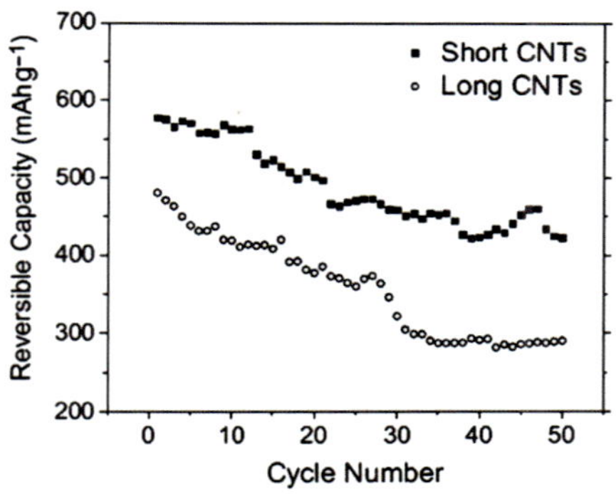

Figure 5.9. Variation of CREV with number of cycles at a current density of 25 mAg^{-1}.

Source: Hui Pan, Creative Commons License.

5.2.2. Si Nanotubes

Si is a highly attractive and competitive option to replace graphite as an anode material. It has a theoretical capacity of up to 4,200 $mAhg^{-1}$ and operates at a low working voltage. Nevertheless, the fluctuation in volume during battery operation can result in significant mechanical harm to the silicon electrodes and lead to a rapid decline in capacity over repeated cycles. An effective strategy involves integrating carbon nanotubes (CNTs) into a silicon (Si) matrix (Yi et al., 2019). In contrast, decreasing the dimensions of silicon materials is anticipated to mitigate the issue of pulverization due to the little absolute local volume change that is projected to occur in nanostructures, particularly nanotubes. Park et al. found that Si nanotubes had a remarkably high reversible charge capacity of 3,247 $mAhg^{-1}$, along with a coulombic efficiency of 89%. Additionally, these nanotubes demonstrated excellent capacity retention even at a 5C rate (= 15 A g^{-1}). According to Yoo et al., the reversible charge capacity may be increased to 3,750 mAh g^{-1}, while the coulombic efficiency can reach 97.5% (Daskalakis et al., 2023). Wu et al. developed a unique anode called a double-walled Si–SiOx nanotube (DWSiNT). This anode consists of an inner wall made of active silicon and an outside wall made of restricting SiO_x. This design enables the passage of lithium ions. The DWSiNT structure effectively resolved the problems related to the stability of silicon material and solid-electrolyte interphase. It demonstrated a prolonged cycle life of 6,000 cycles with 88% capacity retention. Additionally, it exhibited a high specific charge capacity of 2,971/1,780 mAh g^{-1} at 5°C and 940/600 mAh g^{-1} at 12°C, based on the weight of silicon and total-DWSiNT, respectively. Furthermore, it achieved rapid charge and discharge rates of up to 20°C (Yoo et al., 2012).

5.2.3. TiO_2 Nanotube

Titanium dioxide (TiO_2) is a highly promising material for use as an anode in lithium-ion batteries (LIBs) due to its advantageous characteristics. These include a high working voltage (over 1.5 V compared to lithium), as well as structural stability during the process of lithium insertion and extraction. This stability is maintained because TiO_2 experiences minimal volume expansion (only 3%), which prevents the decomposition of the organic electrolyte and the occurrence of short circuits caused by electrode swelling during cycling. Nevertheless, the limited rate capability of TiO_2-based anode hinders its practical application (Salari et al., 2014). Typically, there are two methods used to address the limitation: integrating active/conducting elements into TiO_2, and lowering its size. The porous TiO_2 nanostructures, particularly nanotubes, exhibit high lithium flux and short lithium diffusion length, resulting in enhanced kinetics related to lithium. This is due to the large interface area between the active materials and the electrolyte, as well as the nanometer-scale dimensions. Bi et al. conducted a study where they employed electrochemical anodization procedures to manufacture TiO_2 -NTs on Ti foam. These TiO_2 -NTs were then used as anodes in lithium-ion batteries. The anodes demonstrated high capacities of 103 $mAhcm^{-1}$ at 10 mA cm^{-1} and 83 mAh cm^{-1} at 500 mA cm^{-1}, respectively. They additionally demonstrated that the sample on Titanium foam had superior performance compared to the sample on Titanium foil. Li et al. discovered that the efficiency of TiO_2 nanotubes (NTs) can be enhanced through the utilization of ionic liquids for the anodization of titanium (Ti) (Kim et al., 2009).

SUMMARY

- This chapter provides a comprehensive summary of the current advancements in the use of nanotubes as materials for storing energy. This article systematically discusses the impact of the structures/morphologies, functionalization/doping, complex composites, and experimental methods on the performance of materials in hydrogen storage and lithium battery applications. These topics have gained significant attention from physicists, chemists, materials scientists, and industry professionals.
- The performance of pure NT-based storage cells for energy storage is directly influenced by the physical features of the NTs, such as their specific surface area. This dependence is highly reliant on the synthesis and post-treatment processes used for the NTs. The particular surface area, which may be utilized by H_2 in hydrogen storage and Li^+ in the battery, is a crucial aspect for enhancing performance. Additional parameters, including the size and distribution of pores, the alignment of nanotubes, as well as their length and diameter, significantly impact their storage capabilities. The conductivity of nanotubes (NTs) is crucial for their utilization in supercapacitors and lithium batteries. Optimizing these characteristics can enhance the performance of the storage cell.
- Functionalizing, doping, aligning, and packing nanotubes can enhance their capacity by improving the distribution of pore sizes, conductivity, ion diffusivity, and incorporating functional groups. Out of all the types of nanotubes (NTs), carbon-related nanotubes (CNTs, BCN, BN, g-C_3N_4, etc.) exhibit superior stability due to their exceptional physical and chemical properties. The non-carbon nanotubes (NTs), such as oxide NTs, exhibit a significant capacity but possess lesser stability, recyclability, and conductivity compared to carbon-based NTs, particularly for use in supercapacitors and lithium batteries.
- It is important to note that the materials based on nanotechnology (NT) for energy storage have not yet met the expectations in terms of cost and large-scale production in the industry. However, there have been considerable advancements in NT-based energy storage materials during the past decade. The development of novel materials and composites, such as porous nanotubes (NTs) and their composites, has the potential to enhance energy storage capabilities by increasing performance (including high capacity and dependability) while also reducing costs and simplifying production processes.

MULTIPLE CHOICE QUESTIONS

1. **Which of the following nanotubes are considered promising for hydrogen storage due to their high surface area and unique structural properties?**
   a. Silicon nanotubes
   b. Carbon nanotubes
   c. Titanium dioxide nanotubes
   d. Boron nitride nanotubes
2. **Inorganic nanotubes, such as those made from boron nitride, are explored for hydrogen storage because they:**
   a. Are abundant and inexpensive
   b. Exhibit distinct adsorption properties suitable for hydrogen
   c. Have low thermal conductivity
   d. Lack structural stability
3. **Which composite material is known to enhance hydrogen storage by combining carbon nanotubes with metal oxides?**
   a. CNT/silicon oxide
   b. CNT/boron nitride
   c. CNT/oxide composite
   d. CNT/titanium oxide
4. **Pure carbon nanotubes are significant for lithium-ion batteries because they:**
   a. Have a high volumetric expansion rate
   b. Provide excellent electrical conductivity and mechanical strength
   c. Are inexpensive and readily available
   d. Have low surface area
5. **Silicon nanotubes are particularly noted for their high theoretical capacity for lithium storage, but they face challenges related to:**
   a. Low thermal conductivity
   b. High mechanical strength
   c. Volumetric expansion during lithiation
   d. High cost of production

REVIEW QUESTIONS

1. What are the unique properties of carbon nanotubes (CNTs) that make them suitable for hydrogen storage.
2. How do inorganic nanotubes differ from carbon nanotubes in terms of hydrogen storage capabilities.
3. How do pure carbon nanotubes improve the performance of lithium-ion batteries.
4. What are the advantages and disadvantages of using silicon nanotubes (Si Nanotubes) as anode materials in lithium-ion batteries.
5. Compare and contrast the use of carbon nanotubes and silicon nanotubes in lithium battery technology.

Answers to Multiple Questions

1. (b); 2. (b); 3. (c); 4. (b); 5. (c)

REFERENCES

1. Arya, N., Verma, D., & Balakrishnan, V. (2023). Fabrication of vertically aligned CNT-vanadium oxide hybrid architecture with enhanced compressibility and supercapacitor performance. *Nanotechnology*, *34*(11), 115401.
2. Azam, M. A., Manaf, N. S. A., Talib, E., & Bistamam, M. S. A. (2013). Aligned carbon nanotube from catalytic chemical vapor deposition technique for energy storage device: A review. *Ionics*, *19*, 1455–1476.
3. Bai, Y., Shen, B., Zhang, S., Zhu, Z., Sun, S., Gao, J., & Wei, F. (2019). Storage of mechanical energy based on carbon nanotubes with high energy density and power density. *Advanced Materials*, *31*(9), 1800680.
4. Brzhezinskaya, M., Belenkov, E. A., Greshnyakov, V. A., Yalovega, G. E., & Bashkin, I. O. (2019). New aspects in the study of carbon-hydrogen interaction in hydrogenated carbon nanotubes for energy storage applications. *Journal of Alloys and Compounds*, *792*, 713–720.
5. Cao, Z., & Wei, B. B. (2013). A perspective: Carbon nanotube macro-films for energy storage. *Energy & Environmental Science*, *6*(11), 3183–3201.
6. Chen, X., Zhang, Y., Gao, X. P., Pan, G. L., Jiang, X. Y., Qu, J. Q., & Song, D. Y. (2004). Electrochemical hydrogen storage of carbon nanotubes and carbon nanofibers. *International Journal of Hydrogen Energy*, *29*(7), 743–748.

7. Daskalakis, S., Kostopoulou, A., Brintakis, K., Stratakis, E., Prasadam, V. P., Menguelti, K., & Vernardou, D. (2023). Investigation of Si-Coated Multiwalled Carbon Nanotubes as Potential Electrodes for Multivalent Metal-Ion Electrochemical Energy Storage Systems. *The Journal of Physical Chemistry C*, *127*(27), 13364–13379.

8. Dillon, A. C. (2010). Carbon nanotubes for photoconversion and electrical energy storage. *Chemical Reviews*, *110*(11), 6856–6872.

9. Fan, X., Zhou, H., & Guo, X. (2015). WC nanocrystals grown on vertically aligned carbon nanotubes: An efficient and stable electrocatalyst for hydrogen evolution reaction. *ACS Nano*, *9*(5), 5125–5134.

10. Froudakis, G. E. (2011). Hydrogen storage in nanotubes & nanostructures. *Materials Today*, *14*(7–8), 324–328.

11. Ge, D., Yang, L., Honglawan, A., Li, J., & Yang, S. (2014). In situ synthesis of hybrid aerogels from single-walled carbon nanotubes and polyaniline nanoribbons as free-standing, flexible energy storage electrodes. *Chemistry of Materials*, *26*(4), 1678–1685.

12. Gordon, P. A., & Saeger, R. B. (1999). Molecular modeling of adsorptive energy storage: Hydrogen storage in single-walled carbon nanotubes. *Industrial & Engineering Chemistry Research*, *38*(12), 4647–4655.

13. Itas, Y. S., Razali, R., Tata, S., Idris, A. M., & Khandaker, M. U. (2023). Studies of the hydrogen energy storage potentials of Fe-and Al-doped silicon carbide nanotubes (SiCNTs) by optical adsorption spectra analysis. *Journal of Energy Storage*, *72*, 108534.

14. Jampani, P. H., Kadakia, K., Hong, D. H., Epur, R., Poston, J. A., Manivannan, A., & Kumta, P. N. (2013). CVD derived vanadium oxide nano-sphere-carbon nanotube (CNT) nano-composite hetero-structures: High energy supercapacitors. *Journal of The Electrochemical Society*, *160*(8), A1118.

15. Kapoor, A., Patrike, A. L., Singh, N., Thauer, E., Ottmann, A., Klingeler, R., & Bajpai, A. (2023). Synthetically encapsulated & self-organized transition metal oxide nano-structures inside carbon nanotubes as robust: Li-ion battery anode materials. *Journal of Physics D: Applied Physics*, *56*(42), 425504.

16. Kim, M. S., Lee, T. W., & Park, J. H. (2009). Controlled TiO2 nanotube arrays as an active material for high power energy-storage devices. *Journal of the Electrochemical Society*, *156*(7), A584.

17. Koh, G., Zhang, Y. W., & Pan, H. (2012). First-principles study on hydrogen storage by graphitic carbon nitride nanotubes. *International Journal of Hydrogen Energy*, *37*(5), 4170–4178.

18. Kurzweil, P. (2015). Lithium battery energy storage: State of the art including lithium–air and lithium–sulfur systems. In *Electrochemical Energy Storage for Renewable Sources and Grid Balancing* (Vol. 4, pp. 269–307).
19. Lin, Z., Wang, K., Wang, X., Wang, S., Pan, H., Liu, Y., & Cao, S. (2020). Carbon-coated graphitic carbon nitride nanotubes for supercapacitor applications. *ACS Applied Nano Materials*, *3*(7), 7016–7028.
20. Ma, S. B., Nam, K. W., Yoon, W. S., Yang, X. Q., Ahn, K. Y., Oh, K. H., & Kim, K. B. (2008). Electrochemical properties of manganese oxide coated onto carbon nanotubes for energy-storage applications. *Journal of Power Sources*, *178*(1), 483–489.
21. Mahdizadeh, S. J., & Goharshadi, E. K. (2014). Hydrogen storage on silicon, carbon, and silicon carbide nanotubes: A combined quantum mechanics and grand canonical Monte Carlo simulation study. *International Journal of Hydrogen Energy*, *39*(4), 1719–1731.
22. Martincic, M., Vranic, S., Pach, E., Sandoval, S., Ballesteros, B., Kostarelos, K., & Tobias, G. (2019). Non-cytotoxic carbon nanocapsules synthesized via one-pot filling and end-closing of multi-walled carbon nanotubes. *Carbon*, *141*, 782–793.
23. Masih-Tehrani, M., Ha'iri-Yazdi, M. R., Esfahanian, V., & Safaei, A. (2013). Optimum sizing and optimum energy management of a hybrid energy storage system for lithium battery life improvement. *Journal of Power Sources*, *244*, 2–10.
24. Okoroanyanwu, U., Bhardwaj, A., Einck, V., Ribbe, A., Hu, W., Rodriguez, J. M., & Watkins, J. J. (2021). Rapid preparation and electrochemical energy storage applications of silicon carbide and silicon oxycarbide ceramic/carbon nanocomposites derived via flash photothermal pyrolysis of organosilicon preceramic polymers. *Chemistry of Materials*, *33*(2), 678–694.
25. Rakhi, R. B., Chen, W., Cha, D., & Alshareef, H. N. (2012). Nanostructured ternary electrodes for energy-storage applications. *Advanced Energy Materials*, *2*(3), 381–389.
26. Rashidi, A. M., Nouralishahi, A., Khodadadi, A. A., Mortazavi, Y., Karimi, A., & Kashefi, K. (2010). Modification of single wall carbon nanotubes (SWNT) for hydrogen storage. *International Journal of Hydrogen Energy*, *35*(17), 9489–9495.
27. Salari, M., Aboutalebi, S. H., Chidembo, A. T., Konstantinov, K., & Liu, H. K. (2014). Surface engineering of self-assembled TiO2 nanotube

arrays: A practical route towards energy storage applications. *Journal of Alloys and Compounds*, *586*, 197–201.

28. Samy, M. M., Mohamed, M. G., & Kuo, S. W. (2020). Pyrene-functionalized tetraphenylethylene polybenzoxazine for dispersing single-walled carbon nanotubes and energy storage. *Composites Science and Technology*, *199*, 108360.
29. Shi, K., Ren, M., & Zhitomirsky, I. (2014). Activated carbon-coated carbon nanotubes for energy storage in supercapacitors and capacitive water purification. *ACS Sustainable Chemistry & Engineering*, *2*(5), 1289–1298.
30. Shrivastav, V., Sundriyal, S., Goel, P., Kaur, H., Tuteja, S. K., Vikrant, K., & Deep, A. (2019). Metal-organic frameworks (MOFs) and their composites as electrodes for lithium battery applications: Novel means for alternative energy storage. *Coordination Chemistry Reviews*, *393*, 48–78.
31. Soares, D. M., Vicentini, R., Peterlevitz, A. C., Rodella, C. B., da Silva, L. M., & Zanin, H. (2019). Tungsten oxide and carbide composite synthesized by hot filament chemical deposition as electrodes in a□ueous-based electrochemical capacitors. *Journal of Energy Storage*, *26*, 100905.
32. Sun, L., Wang, X., Wang, Y., & Zhang, Q. (2017). Roles of carbon nanotubes in novel energy storage devices. *Carbon*, *122*, 462–474.
33. Suryanto, B. H., Chen, S., Duan, J., & Zhao, C. (2016). Hydrothermally driven transformation of oxygen functional groups at multiwall carbon nanotubes for improved electrocatalytic applications. *ACS Applied Materials & Interfaces*, *8*(51), 35513–35522.
34. Tang, W., Zhu, Y., Hou, Y., Liu, L., Wu, Y., Loh, K. P., & Zhu, K. (2013). Aqueous rechargeable lithium batteries as an energy storage system of superfast charging. *Energy & Environmental Science*, *6*(7), 2093–2104.
35. Thackeray, M. M., Wolverton, C., & Isaacs, E. D. (2012). Electrical energy storage for transportation—approaching the limits of, and going beyond, lithium-ion batteries. *Energy & Environmental Science*, *5*(7), 7854–7863.
36. Tuci, G., Zafferoni, C., Rossin, A., Milella, A., Luconi, L., Innocenti, M., & Giambastiani, G. (2014). Chemically functionalized carbon nanotubes with pyridine groups as easily tunable N-decorated nanomaterials for the oxygen reduction reaction in alkaline medium. *Chemistry of Materials*, *26*(11), 3460–3470.
37. Uceta, H., Cabrera-Espinoza, A., Barrejón, M., Sánchez, J. G., Gutierrez-Fernandez, E., Kosta, I., & Delgado, J. L. (2023). p-Type Functionalized Carbon Nanohorns and Nanotubes in Perovskite Solar Cells. *ACS Applied Materials & Interfaces*, *15*(38), 45212-–45228.

38. Wang, G. X., Ahn, J. H., Yao, J., Lindsay, M., Liu, H. K., & Dou, S. X. (2003). Preparation and characterization of carbon nanotubes for energy storage. *Journal of Power Sources*, *119*, 16–23.

39. Wen, L., Li, F., & Cheng, H. M. (2016). Carbon nanotubes and graphene for flexible electrochemical energy storage: From materials to devices. *Advanced Materials*, *28*(22), 4306–4337.

40. Xiang, Z., Dai, Q., Chen, J. F., & Dai, L. (2016). Edge functionalization of graphene and two-dimensional covalent organic polymers for energy conversion and storage. *Advanced Materials*, *28*(29), 6253–6261.

41. Yan, L., Xu, Y., Zhou, M., Chen, G., Deng, S., Smirnov, S., & Zou, G. (2015). Porous TiO2 conformal coating on carbon nanotubes as energy storage materials. *Electrochimica Acta*, *169*, 73–81.

42. Yi, Z., Lin, N., Zhao, Y., Wang, W., Qian, Y., Zhu, Y., & Qian, Y. (2019). A flexible micro/nanostructured Si microsphere cross-linked by highly-elastic carbon nanotubes toward enhanced lithium-ion battery anodes. *Energy Storage Materials*, *17*, 93–100.

43. Yoo, J. K., Kim, J., Jung, Y. S., & Kang, K. (2012). Scalable fabrication of silicon nanotubes and their application to energy storage. *Advanced Materials*, *24*(40), 5452–5456.

44. Zhang, R., Wen, Q., Qian, W., Su, D. S., Zhang, Q., & Wei, F. (2011). Superstrong ultralong carbon nanotubes for mechanical energy storage. *Advanced Materials*, *23*(30), 3387–3391.

45. Zhao, Z., Zhu, Z., Wang, F., Li, S., Bao, X., Zhang, L., & Yang, Y. (2021). Bimetallic carbides embedded in heteroatom-doped carbon nanotubes for efficient electrocatalytic hydrogen evolution reaction and high-performance lithium storage. *Chemical Engineering Journal*, *415*, 128885.

46. Zikalala, S. A., Kuvarega, A. T., Mamba, B. B., Mhlanga, S. D., & Nxumalo, E. N. (2018). The effect of synthetic routes on the physicochemical properties and optical response of N-doped titania–Oxidized carbon nanotube nanohybrids. *Materials Today Chemistry*, *10*, 1–18.

CHAPTER 6

NANO COMPOSITES FOR SURFACE COATINGS

LEARNING OBJECTIVES

At the end of this chapter, readers will be able to:

1. Gain insight into the nanocomposites and their classification
2. Comprehend different types of matrix-based nanocomposites
3. Understand the fabrication process of nanocomposites
4. Comprehend the applications of nanocomposites in different fields

INTRODUCTORY EXAMPLE

BASF, one of the world's leading chemical companies, has been at the forefront of developing advanced nanocomposite coatings. These coatings leverage the unique properties of nanoparticles to enhance performance in a variety of applications. For instance, BASF's line of nanocomposite coatings includes products that provide superior corrosion resistance, scratch resistance, and UV protection.

One notable product is the BASF R-M AGILIS, a high-performance automotive coating. This nanocomposite coating incorporates nanoparticles into a polymer matrix to create a surface that is highly resistant to environmental wear and tear. The nanoparticles in the coating interact with the polymer matrix to create a strong, durable film that can withstand harsh conditions, maintaining the vehicle's aesthetic appeal and structural integrity over time.

The use of nanocomposite coatings in the automotive industry exemplifies the practical application and benefits of this technology. By integrating nanoparticles, BASF has created coatings that not only extend the life of automotive components but also enhance their performance. This approach is crucial in a sector where durability, efficiency, and appearance are paramount.

UNIT INTRODUCTION

A nanocomposite is a material that consists of a matrix with the deliberate addition of nanoparticles in order to enhance specific qualities. Nanocomposites are a type of composite material that harnesses the advantageous features of unique materials at the nanoscale (Deyab, 2020). A nanocomposite is a solid material composed of many phases, with at least one, two, or three dimensions measuring less than 100 nm. The utilization of nano-composite material has expanded considerably to include many systems, such as 1-dimensional, 2-dimensional, 3-dimensional, and amorphous materials, which consist of different components and are mixed at the nanoscale level. These materials include nanoparticles in a traditional material matrix. The incorporation of nanoparticles represents a substantial improvement in characteristics like as mechanical robustness, resilience, and electrical or thermal conductance. More broadly, this description encompasses porous media, colloids, gels, and copolymers. However, it is more frequently employed to denote the solid amalgamation of a matrix on a large scale and phases on a nano-sized scale that possess opposing qualities as a result of disparities in structure and chemistry. Nanocomposites occur naturally in a variety of structures, such as the shell of an abalone and bone (Bredeau et al., 2008). The use of materials abundant in nanoparticles has been prevalent for an extended period, predating the scientific exploration of their physical and chemical characteristics. The properties of the nanocomposite, such as electrical, mechanical, thermal, optical, electrochemical, and catalytic capabilities, will exhibit substantial variations compared to those of the individual components. Moreover, it offers the chance to combine multiple attributes that are unattainable in a singular substance, such as versatile mechanical properties and superconductivity. The nanocomposite organic, inorganic compounds are a widely recognized group that is now experiencing tremendous growth in research. The properties of nano-composite materials are determined by both the attributes of their constituents and their structure and interfacial characteristics (Palmero et al., 2014).

This section is centered around the present methodologies and upcoming patterns in nanocomposite materials utilized for surface coatings. This section will not comprehensively address all aspects of such coatings. Nevertheless, its primary objective is to offer a detailed understanding of the advancements achieved so far and the forthcoming trajectory of the domain (Meera et al., 2012).

Remember

Nanocomposite coatings are not just for high-tech applications; they are also used in everyday products like sports equipment, packaging materials, and even cosmetics.

6.1. CLASSIFICATION OF NANOCOMPOSITES

Learning Objectives

- Comprehend the organic/inorganic nanocomposites
- Gain knowledge about lamellar nanocomposites

There are primarily two methods of categorizing nanocomposites. These are nanocomposites consisting of both organic and inorganic materials. The researchers put forth significant effort to achieve dominance over nanostructures through synthetic methods. The properties of the nanocomposites are influenced by both the inherent characteristics of the parent materials and their shape and interfacial properties. Nanocomposites can be categorized based on the type of inorganic components they contain. These components can include three-dimensional structures like zeolites, two-dimensional layered materials such as ceramics, metal phosphates, metal oxides, and chalcogenides, as well as a single-dimensional and zero-dimensional materials like chains and clusters of $(Mo_3Se_3\text{-})n$ (Sharma et al., 2021).

6.1.1. Organic/Inorganic Nanocomposites

The organic/inorganic nanocomposites are relevant to bio-ceramics and biomineralization, particularly in terms of in-situ development and polymerization of biopolymers and inorganic matrices (Kango et al., 2013).

6.1.2. Lamellar Nanocomposites

Lamellar nanocomposites are a type of composite where the interactions of the two phases at the interface are maximized. Lamellar nanocomposites may be categorized into two distinct classes: intercalated and exfoliated. They have clearly defined and organized intralamellar space that can potentially be accessed by foreign organisms. By acting as matrix or hosts for polymers, they have the capability to produce intriguing hybrid nanocomposite materials. In the previous scenario, the polymer chains and inorganic layers are arranged in a consistent ratio, and there is a particular amount of polymer layers in the space between the inorganic layers. There is a wide range of inorganic layered materials. The intercalated nanocomposites have a compound-like nature due to the polymer/layer ratio being stable. These materials are noteworthy for their electrical and charge transport capabilities. In exfoliated nanocomposites, the distance between the layers is consistently greater than 100 Å, and the number of polymer chains among the layers can be adjusted continually. Conversely, exfoliated nanocomposites are particularly intriguing due to their exceptional mechanical characteristics (Patil et al., 2005).

6.2. MATRIX BASED NANOCOMPOSITES

Learning Objectives

- Understand the processing and preparation of polymer matrix nanocomposites
- Comprehend the metal matrix nanocomposites
- Understand the preparation of carbon matrix nanocomposites

A nanocomposite is a material that consists of many solid phases, such as metal, ceramic, or polymer, with at least one dimension falling inside the nanoscale range, either compositionally or structurally (Dhatarwal et al., 2021).

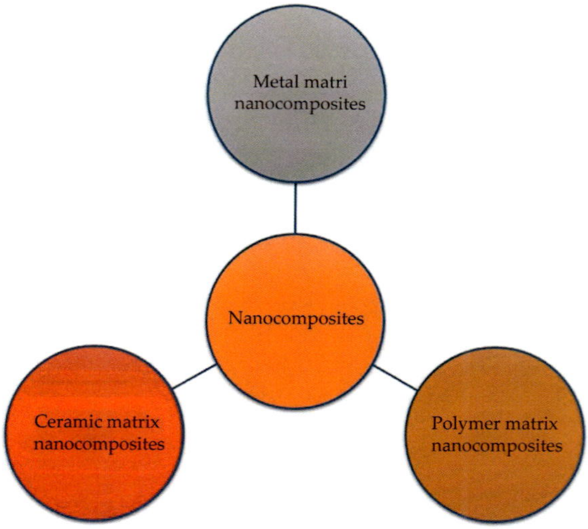

Figure 6.1. *Classification of matrix-based nanocomposites.*

Source: Fouad Damiri, Creative Commons License.

6.2.1. Polymer Matrix Nanocomposite

Polymer matrix composites can be processed at much lower temperatures compared to other composites. There are three methods for manufacturing polymer nanocomposites, which depend on the types of polymer matrices: conventional composite, intercalated composite, and exfoliated nanocomposite (Figure 6.1). Polymer matrix composites can be classified into two primary categories: thermosetting and thermoplastic composites. Polymer nanocomposites provide a novel alternative to conventionally filled polymers. These materials exhibit important properties such as non-linear spectrum characteristics, electronic conductivity, and luminescence (Hamming et al., 2009). The material possesses a wide range of applications, including chemical sensors, electroluminescent devices, electrocatalytic systems, energy storage units, intelligent windows, and memory technologies. The study focused on investigating the potential of a nanocomposite consisting of different polymeric materials, such as polystyrene-coacrylonitrile, polystyrene, polypyrrole, polyacrylates, polyaniline, polysulfone, polyimide, and epoxy, in combination with both unmodified and modified clays, for the development of corrosion-resistant coatings on metal surfaces. Moreover, it is frequently used in the production of items such as packing sheets and tires. These innovative materials are specifically engineered to function as affordable substitutes for expensive technological components including gear systems in timber drilling machines, wear-resistant materials, and the production of plastic film barriers for the food industry. The coatings of magnetic particles are highly fascinating due to their substantial applications, including technological energy conversion, magnetic recording, magnetic fluids, and magnetic refrigeration systems (Sinsawat et al., 2003) (Figure 6.2).

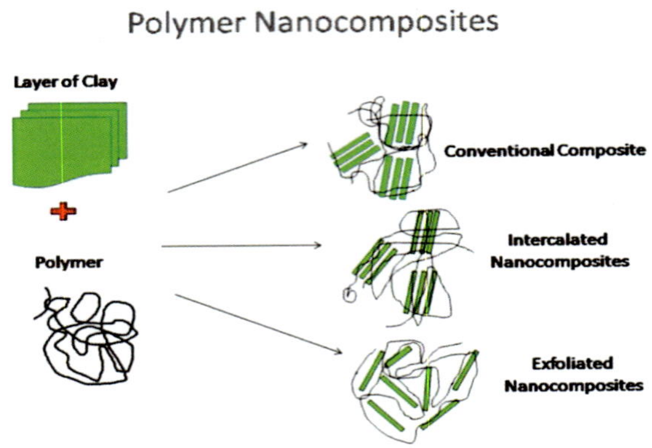

Figure 6.2. Different types of polymer nanocomposites.

Source: Shivani Soni, Creative Commons License.

6.2.2. Metal Matrix Nanocomposite

Metal matrix nanocomposites can also be referred to as reinforced metal matrix composites. These composites can be classified as either continuous or discontinuous reinforced materials. Carbon nanotube metal matrix composites are a significant class of nanocomposites now under development to harness the remarkable elasticity and conductivity properties of carbon nanotube materials (Zhang & Chen, 2006). The application of the nano composite coating technology entails the incorporation of nano-sized oxide particles into metal matrix coatings, leading to remarkable hardness, noteworthy corrosion resistance, and superior wear resistance. Moreover, it possesses the potential to generate very opaque coatings suitable for many applications, such as solar thermal and water desalination fields. There is no need for any extravagant equipment and it may be enlarged to achieve industrial-scale manufacturing. This technology has also enabled inexperienced researchers to perform more extensive investigations on diverse metal matrix and nano coatings, in combination with various substrates. The substrates are composed of steel, aluminum, magnesium, and copper alloys, while the nanocoatings are composed of oxides, carbides, silicates, nitrides, and other similar compounds. The coating properties that can be improved include hardness, durability, resistance to oxidation and corrosion, as well as various physical characteristics such as optical, electrical, and magnetic capabilities (Ying & Zhang, 2000). The forefront research areas in metal matrix nanocomposites encompass boron nitride reinforced metal matrix composites, carbon nanotube metal matrix composites, and carbon nitride metal matrix composites. An example of an innovative nanocomposite is the energetic nanocomposite, typically consisting of a hybrid sol-gel with a silica base. By combining it with metal oxides and nano powdered aluminum, it possesses the capability to generate superthermite substances. The utilization of synthetic dielectrics holds the promise of surpassing the dielectric constraints of polymers by creating advanced polymer composites. Metal-filled polymer composites possess the ability to meet the demands of upcoming dielectric devices, such as embedded capacitors with significant capacitance densities (Saheb et al., 2012).

6.2.3. Carbon Matrix Nanocomposites

Carbon is quite a remarkable element; it exists in four different forms: diamond, graphite, carbynes, and fullerenes. Each of these forms has important applications in science and technology. Highly organized graphite fibers inserted in a carbon matrix make up carbon-carbon composites. Of all ceramic composite materials, carbon-carbon composites are the most versatile and have a wide range of uses in many industries (Tjong, 2013). Fibers oriented in different directions, as well as hydrocarbons and carbonaceous polymers as matrix precursors, are used to make these composites. They are created by repeatedly accumulating a carbon matrix on a fiber using chemical vapor deposition, pyrolysis, and impregnation processes. In comparison to steel or other metals, carbon-carbon composites are often stronger, stiffer, and lighter. The process of C-C composite processing involves encircling the graphite fibers with a carbon matrix (Moghadam et al., 2015).

There are two common methods for creating the matrix: applying a resin and using chemical vapor deposition. The thermal characteristics of carbon-carbon composites are the most important class of attributes. Due to their strong thermal conductivity and extremely small thermal expansion coefficients, carbon-carbon composites are dimensionally stable over a broad temperature range. Mechanical characteristics of carbon-carbon composites are retained at temperatures higher than 2000°C. Additional characteristics include resistance to chemical and biological corrosion, low weight, good electrical conductivity, high abrasion resistance, low hygroscopicity, and non-brittle failure. Carbon-carbon composites are very practical and may be shaped into intricate designs (Lin et al., 2003).

Did You Know?

Some nanocomposite coatings can make surfaces self-cleaning. These coatings use nanoparticles that cause water to bead up and roll off, taking dirt and grime with it. This technology is inspired by the lotus leaf, known for its self-cleaning properties.

6.3. FABRICATION OF NANOCOMPOSITES

Learning Objectives

- Understand the fabrication process of nanocomposites
- Comprehend the process of melt processing

Several techniques can be employed to create nanocomposites of conducting polymers, such as in situ polymerization with nanoparticles, coating inorganic polymers, electrochemical encapsulation, and colloidal dispersions. These techniques have created fresh possibilities for the production of materials. A more thorough comprehension is required to grasp the relationship between the formulation, structure, and properties of platelets and their tendency to flake and disperse. The solution-induced intercalation approach entails dissolving the polymer in an organic solvent. Afterward, the clay is disseminated in the solution that is produced, and either the solvent is completely used up or the polymer forms solid particles. This approach leads to insufficient dispersion of clay particles, along with additional problems such as the expensive nature of solvents, the need for a large amount of solvent to obtain a noticeable distribution of fillers, the possibility of technical phase separation, and concerns regarding health and safety. The solvent route strategy is used for water-soluble polymers due to clear and logical reasons (Huang et al., 2021).

Nanocomposite coatings consist of nanocrystalline phases embedded in an amorphous matrix. These coatings provide substantial enhancements in abrasion resistance, hardness, toughness, and friction reduction. A nanocomposite polyhedral oligomeric silsesquioxane-poly urethane material, specifically a POSS-PCU48, has been created (Sundarrajan & Ramakrishna, 2007). This material exhibits ideal mechanical and physiological properties for cardiovascular applications, including a natural antithrombotic effect. This polymer has been implanted in humans for applications such as vascular bypass grafts, lacrimal ducts, and tracheal implants. Previously, a range of fabrication techni□ues such as thermoforming, cavity molding, injection molding, film fabrication, and dip coating were utilized for the production of polymer nanocomposites (Kim et al., 2020) (Figure 6.3).

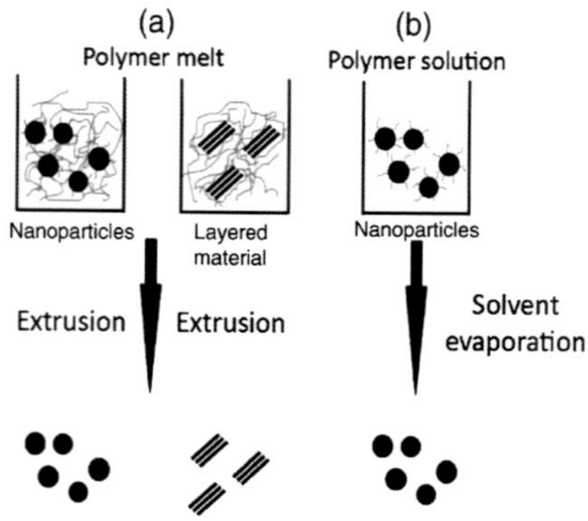

Figure 6.3. *Illustration of preparation methods for nanocomposites.*

Source: Muhammet Toprak, Creative Commons License.

6.3.1. Melt Processing

This process involves the direct dispersion of silicate layers into the polymers while they are in a molten state. Prior to utilizing this technique, it is necessary to subject the silicates to surface treatment through organ modification, as previously indicated. In situ polymerization involves dispersing clay layers into a matrix through polymerization. This is achieved by combining the silicate layers with the monomer, along with the polymerization initiator and catalyst. The process involves polymerizing a monomer within the host material, using the host material itself as the oxidizing agent (Ni et al., 2023).

Intercalative polymerization occurs when host matrices with a strong attraction for electrons cause the oxidative polymerization of suitable monomers within their structure. The process involves the insertion of monomers between the layers of a material, followed by a reaction that converts them into a solid polymer inside the same layers (Yu et al., 2019). This pathway facilitates the formation of conjugated polymers within hosts that do not undergo oxidation. The process of directly enclosing polymer chains within single layers of a host material that are disseminated as colloids. This method provides the opportunity to utilize a wide range of nano-composites containing various types of polymers and hosts (Rane et al., 2018).

Practice Problem

Describe the sol-gel process for fabricating ceramic matrix nanocomposites and explain how this method ensures uniform nanoparticle distribution within the matrix.

Solution to Practice Problem

The sol-gel process involves the transition of a colloidal suspension (sol) into a solid network (gel). The steps are as follows.

i. **Precursor Preparation**: Metal alkoxides or metal salts are dissolved in a solvent to form a solution

ii. **Hydrolysis and Condensation**: The solution undergoes hydrolysis and condensation reactions, forming a colloidal suspension of nanoparticles

iii. **Gelation**: The sol transitions into a gel as the network of nanoparticles grows and interconnects

iv. **Drying and Heat Treatment:** The gel is dried to remove the solvent, followed by heat treatment to densify and crystallize the ceramic matrix.

Uniform Distribution: The sol-gel process ensures uniform nanoparticle distribution by forming nanoparticles in situ within the sol, leading to homogeneous incorporation into the matrix during gelation.

6.4. APPLICATIONS OF NANOCOMPOSITES

Learning Objectives

- Comprehend the diverse applications of nanocomposites

The widespread utilization of polymer materials in modern times is due to their exceptional amalgamation of characteristics, lightweight nature, and effortless manipulation throughout manufacturing. However, in order to enhance certain features like as thermodynamic and mechanical stability, a significant amount of additives are incorporated into the polymeric matrix, resulting in the formation of a polymer matrix composite. A composite is a compound formed by combining two or more materials that have distinct physical and chemical properties and may be differentiated by their interface. Composite materials provide a diverse and extensive array of applications (Ates et al., 2020). Composites have several significant advantages compared to many metal compounds, including excellent specific stiffness and strength, excellent durability, resistance to corrosion, low density, and thermal insulation. Nanocomposites offer potential for novel applications across various domains, including mechanically enhanced lightweight parts, non-linear optical technology, battery cathodes and ionics, sensors, nano-wires, and other systems. Nanocomposites provide the opportunity to combine a variety of qualities that would be unattainable in a single material, such as the combination of flexible mechanical properties with superconducting properties (Camargo et al., 2009) (Figure 6.4).

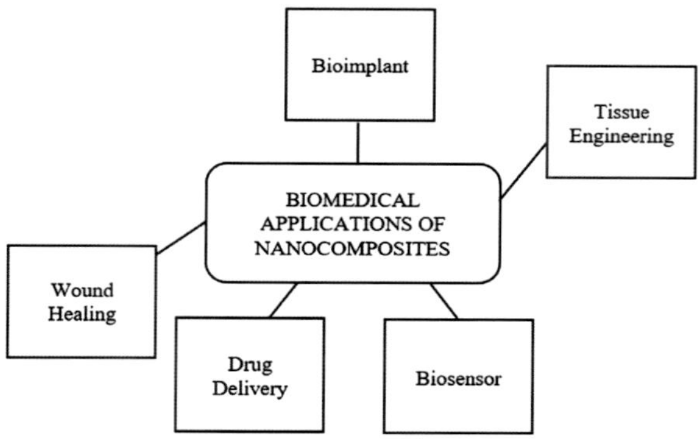

Figure 6.4. *Applications of nanocomposite in biomedical field.*

Source: P. Gopal, Creative Commons License.

6.4.1. Nanocomposites Coatings

The development of the nanocomposite surface coating technique may be based on the existing classic composite coating technology. This approach can be applied in the treatment of plastic injection mold cavity surfaces. The nanocomposite coating technology can be compared to the conventional nickel electro-brush coating and the fundamental method of achieving nanoscale scale on the mold cavity. Nanocomposite materials in the modern day are developed through the processes of manufacture, surface modification, mixing, and shape of various materials. Optically active nanocomposite substances will be an innovative approach, utilizing technology and approaches obtained from other domains of materials (Hauert & Patscheider, 2000). Thermal spraying has sparked more research into the impact of various components on the boundary characteristics of composite coatings. The technology has the capability to produce composite materials at the nano-scale, which are versatile and can be used in the development of advanced materials and technologies for rechargeable batteries, metals, ceramics, fuel cells, cosmetics, superconductors, and medicines. Nanocomposite tribological coatings provide the ability to adapt their surface properties in response to varying environmental circumstances, ensuring optimal tribological performance in any given environment. To address temperature fluctuations, researchers devised nanocomposite coatings consisting of yttria-stabilized zirconia embedded in a gold matrix. These coatings also contained enclosed nanosized pools of MoS_2 and diamond-like carbon. The process of creating coatings involves the utilization of both laser treatment and magnetron sputtering techniques (Gan & Berndt, 2015).

6.4.2. Polymer Nanofibers

Polymer fibers are often manufactured on a large scale by extrusion. This involves pumping a polymer melt or solution via cylindrical dies and spinning it with a take-up mechanism. The resultant fibers often have sizes in the range of 10 micrometers or larger. Electrospinning is the most prominent method of polymer processing currently accessible. Electrospinning, co-electrospinning, and related modeling approaches involving nanofibers produce nano-sized objects (Chronakis, 2005).

6.4.3. Proteins Immobilization

The immobilization of biological items such as proteins, bacteria, or viruses is necessary for various scientific applications, including chromatography, catalysis, sensorics, optical information technology, and the delivery of drugs. This reasoning employs carbon nanotubes, gold particles, and synthetic polymers. The immobilization has primarily been accomplished through adsorption or chemical binding, with a smaller degree of integration of these elements as visitors in host matrices. Polymers are an ideal option for immobilizing biological substances because of the extensive range of synthetic and natural polymers available, as well as the sophisticated techniques utilized to transform them into nanofibers, rods, tubes, and other forms (Wu et al., 2007). Polymer nanostructured fibers, core-shell fibers, hollow fibers, nanotubes, and nanorods provide a flexible foundation for a wide range of applications in both material science and biological science. Core-shell fibers, consisting of nanoparticles with fluid cores and hard shells, can effectively trap proteins, viruses, or bacteria while preserving their functions. These implications can be employed for biosensor applications, as well as for other purposes. Presently, a few probable instances encompass biosensors, tissue engineering, medicine distribution, and enzymatic catalysis. By immobilizing green fluorescent protein (GFP) in nanostructured fibers, a sensor protein is created with a large surface area and short distances for the analyte to reach. A highly efficient biosensor was successfully created to specifically identify different dengue serotypes. This biosensor employs a gold electrode that has been altered and coupled with hybrid organic-inorganic nanocomposites and extracts derived from the Bauhinia species. It has the ability to precisely diagnose dengue infections (Chaudhary et al., 2021).

6.4.4. Carbon Fiber Reinforced Polymer

Carbon fiber-reinforced polymer composite components are susceptible to erosion caused by sand and moisture, as well as damage from sharp things. Utilizing nanoparticles in coating compositions can result in the development of nanocomposite coatings that exhibit high resistance to wear. Researchers are currently working on the development of innovative polymer coating systems that incorporate small particles. These coatings have the potential to significantly improve erosion resistance. The researchers are specifically investigating how these coatings might be applied to composite rotors and blades. These nanocomposite coatings may be customized to possess excellent conductive qualities, both thermally and electrically. This makes them suitable for potential uses in lightning strike protection and de-icing (Ozdemir et al., 2016).

6.4.5. Nanocomposite-Based Nanogenerator

A newly created nanocomposite-based nanogenerator effectively overcomes the significant limitations seen in previous nanogenerators and establishes a straightforward, inexpensive, and extensive self-sustaining energy system. The utilization of piezoelectric effects in nano-generator technology allows for the conversion of nonpolluting energy sources, such as wind and wave vibrations, into a continuous supply of electrical energy. This technology has garnered significant attention in the field of next-generation power generation (Rana et al., 2021).

6.4.6. Antibacterial Nanocomposite

The hybrid sol-gel process can be utilized for the manufacture of antibacterial nanocomposites. This process is specifically developed to apply a protective layer on the surfaces of plastic, metal, and ceramic substrates in order to enhance their antibacterial characteristics. It provides an antibacterial coating to surfaces by applying a layer that eliminates and prevents the growth of bacteria and microorganisms. Furthermore, the antibacterial properties of the nanocomposite coating do not require any external energy, such as UV radiation, to be activated (Rajabi et al., 2018). The nanocomposite is created by combining a polymeric matrix with oxide chalcogenide nanoparticles from the heavy metal oxide class, using the prohibited oxide nanomaterials synthesis technique and process methodology. It can be used to clarify the process of coating surfaces either by immersion or by spraying. It can also be formed into pellets and granules for further processing, such as injection molding. The effectiveness of the ZnO-TiO_2 sol-gel in killing E. Coli and S. aureus bacteria was demonstrated in a study. The antibacterial nanocomposite was found to destroy over 90% of the germs when tested on a concentrated mixture of bacteria. Silver ions interact with the bacterial cell in a harmful manner in three important locations: (1) the peptidoglycan cell wall, leading to cell lysis, (2) bacterial DNA, hindering DNA replication, and (3) bacterial proteins, impeding vital cellular processes like respiration (Hashim et al., 2018).

6.4.7. Semiconductor Nanocrystals

Extensive research has been conducted on semiconductor nanocrystals because of their distinct optical and electrical characteristics, such as a sharp emission band that can be excited across a wide range and their high resistance to photobleaching. These materials are highly intriguing due to their similar dimensions to biomacromolecules and potential applications in biomedical imaging and biological conjugation (Jung et al., 2014). A unique water-soluble hybrid material is being produced, which comprises magnetic nanoparticles and quantum dots enclosed within a silica shell. Superparamagnetic nanoparticles have been researched for their potential use in biomedical applications, including magnetic immobilization, enhancing contrast in MRI scans, and targeting drugs. These nanoparticles might benefit from having a silica shell in order to enhance their wettability and biocompatibility for various applications (Liang et al., 2021).

6.5. FUTURE RESEARCH DIRECTIONS

Learning Objectives

- Comprehend the future research endevors in the field of nanocomposite films and coatings.

The development of nanocomposite surface coatings is still in its early stages. The focus of research in the domain of nanocomposite coatings and films should be on addressing the challenges associated with developing new superior coatings with excellent physical and functional qualities. The objectives include enhancing the fracture toughness of coatings, controlling the grain size in film growth, developing new technology for producing nanocomposite coatings under different chemical and/or physical conditions, increasing the use of coatings in aerospace and automobile manufacturing, improving defensive coatings with resistance to oxidation above 2,000°C, advancing hybrid coatings using nanophase biomaterials for medical use, and achieving nanocrystallization of amorphous materials (Iroegbu & Ray, 2022).

SUMMARY

- To summarize, significant progress has been achieved in the fabrication of nanocomposites. Polymer-based nanocomposites are currently leading the way in applications due to their advanced level of improvement compared to ceramic and metal counterparts, as well as their distinct features. Nanocomposites have shown promising performance in terms of openings and returns, which has generated significant worldwide curiosity about these materials.
- While significant progress has been achieved in comprehending the composition, manipulation, and characteristics of nanocomposite compounds, there is a vast potential for further investigation and use due to the material's exceptional flexibility.
- The potentially detrimental characteristics of nanotechnology products, including their large surface area, crystalline structure, and reactivity, can make it easier for them to enter the environment and interact with cellular components. This can lead to an increase in the negative effects associated with their composition.
- The release of nanoparticles into the atmosphere is a significant concern for health and safety. Within the realm of nanocomposites, numerous unresolved matters persist that require rigorous theoretical analysis and empirical validation. Therefore, there is a growing need for study on the emission of nanoparticles.
- Finally, it is important to address the social implications of this growing and promising technology, particularly in terms of safety when working with nanosized particles with their composites. Nanocomposite thin films and coatings, which are defined by the presence of nanocrystalline grains of transition-metal nitride or carbide within an amorphous phase, show great potential for enhancing the properties of different functional surfaces and are currently a popular focus in surface engineering.

MULTIPLE CHOICE QUESTIONS

1. Which of the following is NOT a common type of nanocomposite matrix?

a. Polymer matrix

b. Ceramic matrix

c. Metal matrix

d. Glass matrix

2. **Polymer matrix nanocomposites (PMNCs) are often enhanced with which of the following nanoparticles?**
   a. Clay
   b. Silicon carbide
   c. Aluminum oxide
   d. Titanium dioxide
3. **What is the primary benefit of using ceramic matrix nanocomposites (CMNCs) in high-temperature applications?**
   a. Enhanced electrical conductivity
   b. Improved thermal stability
   c. Increased flexibility
   d. Reduced weight
4. **Which fabrication technique involves the transition of a solution into a solid gel phase?**
   a. In-situ polymerization
   b. Sol-gel process
   c. Mechanical alloying
   d. Layer-by-layer assembly
5. **In metal matrix nanocomposites (MMNCs), nanoparticles are primarily used to enhance:**
   a. Color and appearance
   b. Electrical conductivity
   c. Strength and corrosion resistance
   d. Thermal expansion
6. **Which of the following is a common application of nanocomposite coatings in the automotive industry?****
   a. UV resistance
   b. Antibacterial properties
   c. Wear resistance
   d. Electrical insulation

REVIEW QUESTIONS

1. What are the main types of nanocomposite matrices? Provide examples for each type?

2. How do polymer matrix nanocomposites (PMNCs) differ from ceramic matrix nanocomposites (CMNCs) in terms of applications and properties?
3. Explain how the choice of matrix material influences the overall properties of nanocomposites?
4. What are the benefits and limitations of mechanical alloying in the synthesis of metal matrix nanocomposites?
5. Identify the main challenges in the development of sustainable and environmentally friendly nanocomposites.

Answers to Multiple Questions

1. (d); 2. (a); 3. (b); 4. (b); 5. (c); 6. (c)

REFERENCES

1. Ates, B., Koytepe, S., Ulu, A., Gurses, C., & Thakur, V. K. (2020). Chemistry, structures, and advanced applications of nanocomposites from biorenewable resources. *Chemical Reviews*, *120*(17), 9304–9362.
2. Bredeau, S., Peeterbroeck, S., Bonduel, D., Alexandre, M., & Dubois, P. (2008). From carbon nanotube coatings to high-performance polymer nanocomposites. *Polymer International*, *57*(4), 547–553.
3. Camargo, P. H. C., Satyanarayana, K. G., & Wypych, F. (2009). Nanocomposites: Synthesis, structure, properties and new application opportunities. *Materials Research*, *12*, 1–39.
4. Chaudhary, K., Kumar, K., Venkatesu, P., & Masram, D. T. (2021). Protein immobilization on graphene oxide or reduced graphene oxide surface and their applications: Influence over activity, structural and thermal stability of protein. *Advances in Colloid and Interface Science*, *289*, 102367.
5. Chronakis, I. S. (2005). Novel nanocomposites and nanoceramics based on polymer nanofibers using electrospinning process—A review. *Journal of Materials Processing Technology*, *167*(2–3), 283–293.
6. Deyab, M. A. (2020). Anticorrosion properties of nanocomposites coatings: A critical review. *Journal of Molecular Liquids*, *313*, 113533.
7. Dhatarwal, P., Choudhary, S., & Sengwa, R. J. (2021). Dielectric and optical properties of alumina and silica nanoparticles dispersed poly (methyl methacrylate) matrix-based nanocomposites for advanced polymer technologies. *Journal of Polymer Research*, *28*(2), 63.
8. Gan, J. A., & Berndt, C. C. (2015). Nanocomposite coatings: Thermal

spray processing, microstructure and performance. *International Materials Reviews*, *60*(4), 195–244.

9. Hamming, L. M., Qiao, R., Messersmith, P. B., & Brinson, L. C. (2009). Effects of dispersion and interfacial modification on the macroscale properties of TiO_2 polymer–matrix nanocomposites. *Composites Science and Technology*, *69*(11–12), 1880–1886.
10. Hashim, A., Agool, I. R., & Kadhim, K. J. (2018). Modern developments in polymer nanocomposites for antibacterial and antimicrobial applications: A review. *Journal of Bionanoscience*, *12*(5), 608–613.
11. Hauert, R., & Patscheider, J. (2000). From alloying to nanocomposites—Improved performance of hard coatings. *Advanced Engineering Materials*, *2*(5), 247–259.
12. Huang, Y., Chen, M., Xie, A., Wang, Y., & Xu, X. (2021). Recent advances in design and fabrication of nanocomposites for electromagnetic wave shielding and absorbing. *Materials*, *14*(15), 4148.
13. Iroegbu, A. O. C., & Ray, S. S. (2022). Recent developments and future perspectives of bio renewable nanocomposites for advanced applications. *Nanotechnology Reviews*, *11*(1), 1696–1721.
14. Jung, J., Yoon, Y. J., He, M., & Lin, Z. (2014). Organic-inorganic nanocomposites composed of conjugated polymers and semiconductor nanocrystals for photovoltaics. *Journal of Polymer Science Part B: Polymer Physics*, *52*(24), 1641–1660.
15. Kango, S., Kalia, S., Celli, A., Njuguna, J., Habibi, Y., & Kumar, R. (2013). Surface modification of inorganic nanoparticles for development of organic–Inorganic nanocomposites—A review. *Progress in Polymer Science*, *38*(8), 1232–1261.
16. Kim, H. J., Lee, J. S., Park, J. M., Lee, S., Hong, S. J., Park, J. S., & Park, K. H. (2020). Fabrication of nanocomposites complexed with gold nanoparticles on polyaniline and application to their nerve regeneration. *ACS Applied Materials & Interfaces*, *12*(27), 30750–30760.
17. Liang, S., Zhang, M., Biesold, G. M., Choi, W., He, Y., Li, Z., & Lin, Z. (2021). Recent advances in synthesis, properties, and applications of metal halide perovskite nanocrystals/polymer nanocomposites. *Advanced Materials*, *33*(50), 2005888.
18. Lin, Y., Zhou, B., Shiral Fernando, K. A., Liu, P., Allard, L. F., & Sun, Y. P. (2003). Polymeric carbon nanocomposites from carbon nanotubes functionalized with matrix polymer. *Macromolecules*, *36*(19), 7199-–7204.

19. Meera, K. M. S., Sankar, R. M., Murali, A., Jaisankar, S. N., & Mandal, A. B. (2012). Sol–gel network silica/modified montmorillonite clay hybrid nanocomposites for hydrophobic surface coatings. *Colloids and Surfaces B: Biointerfaces*, *90*, 204–210.

20. Moghadam, A. D., Omrani, E., Menezes, P. L., & Rohatgi, P. K. (2015). Mechanical and tribological properties of self-lubricating metal matrix nanocomposites reinforced by carbon nanotubes (CNTs) and graphene—A review. *Composites Part B: Engineering*, *77*, 402–420.

21. Ni, J., Zhang, Y., Song, Z., Zhang, P., Cao, Y., Yang, Y., & Wang, J. (2023). Salt-template-assisted melt-processed porous poly (vinylidene fluoride) nanocomposites for highly efficient all-day passive radiative cooling. *Composites Part A: Applied Science and Manufacturing*, *164*, 107311.

22. Ozdemir, N. G., Zhang, T., Aspin, I., Scarpa, F., Hadavinia, H., & Song, Y. (2016). Toughening of carbon fiber reinforced polymer composites with nano rubber for advanced industrial applications. *Express Polymer Letters*, *10*(5), 394–407.

23. Palmero, P., Montanaro, L., Reveron, H., & Chevalier, J. (2014). Surface coating of oxide powders: A new synthesis method to process biomedical grade nano-composites. *Materials*, *7*(7), 5012–5037.

24. Patil, A. J., Muthusamy, E., & Mann, S. (2005). Fabrication of functional protein–organoclay lamellar nanocomposites by biomolecule-induced assembly of exfoliated aminopropyl-functionalized magnesium phyllosilicates. *Journal of Materials Chemistry*, *15*(35–36), 3838–3843.

25. Rajabi, A., Ghazali, M. J., Mahmoudi, E., Azizkhani, S., Sulaiman, N. H., Mohammad, A. W., & Naicker, A. S. (2018). Development and antibacterial application of nanocomposites: Effects of molar ratio on Ag_2O–CuO nanocomposite synthesized via the microwave-assisted route. *Ceramics International*, *44*(17), 21591–21598.

26. Rana, S. S., Zahed, M. A., Rahman, M. T., Salauddin, M., Lee, S. H., Park, C., & Park, J. Y. (2021). Cobalt-nanoporous carbon functionalized nanocomposite-based triboelectric nanogenerator for contactless and sustainable self-powered sensor systems. *Advanced Functional Materials*, *31*(52), 2105110.

27. Rane, A. V., Kanny, K., Abitha, V. K., & Thomas, S. (2018). Methods for synthesis of nanoparticles and fabrication of nanocomposites. In *Synthesis of Inorganic Nanomaterials* (Vol. 3, pp. 121–139). Woodhead Publishing.

28. Saheb, N., Iqbal, Z., Khalil, A., Hakeem, A. S., Al Aqeeli, N., Laoui, T., & Kirchner, R. (2012). Spark plasma sintering of metals and metal matrix nanocomposites: A review. *Journal of Nanomaterials*, *2012*, 18–18.

29. Sharma, A. K., Priya, & Kaith, B. S. (2021). Polymer nanocomposite matrices: Classification, synthesis methods, and applications. In *Handbook of Polymer and Ceramic Nanotechnology* (Vol. 3, pp. 403–428). Cham: Springer International Publishing.

30. Sinsawat, A., Anderson, K. L., Vaia, R. A., & Farmer, B. L. (2003). Influence of polymer matrix composition and architecture on polymer nanocomposite formation: Coarse-grained molecular dynamics simulation. *Journal of Polymer Science Part B: Polymer Physics*, *41*(24), 3272–3284.

31. Sundarrajan, S., & Ramakrishna, S. (2007). Fabrication of nanocomposite membranes from nanofibers and nanoparticles for protection against chemical warfare stimulants. *Journal of Materials Science*, *42*, 8400–8407.

32. Tjong, S. C. (2013). Recent progress in the development and properties of novel metal matrix nanocomposites reinforced with carbon nanotubes and graphene nanosheets. *Materials Science and Engineering: R: Reports*, *74*(10), 281–350.

33. Wu, S., Ju, H. X., & Liu, Y. (2007). Conductive mesocellular silica–carbon nanocomposite foams for immobilization, direct electrochemistry, and biosensing of proteins. *Advanced Functional Materials*, *17*(4), 585–592.

34. Ying, D. Y., & Zhang, D. L. (2000). Processing of Cu–Al_2O_3 metal matrix nanocomposite materials by using high energy ball milling. *Materials Science and Engineering: A*, *286*(1), 152–156.

35. Yu, W. H., Sing, S. L., Chua, C. K., Kuo, C. N., & Tian, X. L. (2019). Particle-reinforced metal matrix nanocomposites fabricated by selective laser melting: A state-of-the-art review. *Progress in Materials Science*, *104*, 330–379.

36. Zhang, Z., & Chen, D. L. (2006). Consideration of Orowan strengthening effect in particulate-reinforced metal matrix nanocomposites: A model for predicting their yield strength. *Scripta Materialia*, *54*(7), 1321–1326.

CHAPTER 7

CARBON NANOTUBES IN NANOMEDICINE

LEARNING OBJECTIVES

At the end of this chapter, readers will be able to:

1. Comprehend the structural and chemical properties of carbon nanotubes
2. Understand the preparation and functionalization of carbon nanotubes
3. Comprehend the applications of carbon nanotubes in medicine
4. Identify the potential toxicity issues associated with CNTs when used in medical applications

Introductory Example

Nanobiotix, a pioneering company in the field of nanomedicine, has been at the forefront of utilizing nanotechnology for cancer treatment. One of their innovative approaches involves the use of carbon nanotubes (CNTs) to enhance the delivery and efficacy of cancer therapeutics.

Nanobiotix employs both covalent and non-covalent functionalization methods to modify CNTs for medical use. Covalent functionalization involves attaching functional groups to the CNTs' surface, which improves their solubility and biocompatibility. Non-covalent functionalization, on the other hand, involves coating CNTs with biocompatible molecules such as polyethylene glycol (PEG) to prevent aggregation and enhance their stability in biological environments.

Nanobiotix has developed CNT-based drug delivery systems that can carry chemotherapy drugs directly to tumor cells. By attaching the drugs to the functionalized CNTs, they achieve targeted delivery, which increases the drug's concentration at the tumor site while minimizing side effects on healthy tissues. In addition to drug delivery, Nanobiotix is exploring the use of CNTs for gene therapy. Functionalized CNTs can carry genetic material such as DNA or RNA into cells, offering a potential method for treating genetic disorders or enhancing the immune response against cancer.

Nanobiotix's work with CNTs exemplifies the transformative potential of nanotechnology in medicine. Their research and development efforts are paving the way for more effective and targeted therapies, particularly in oncology. As they continue to refine their technologies and address safety concerns, the use of CNTs in nanomedicine could revolutionize the treatment of various diseases, offering new hope for patients worldwide.

UNIT INTRODUCTION

Since their discovery by Iijima in 1991, Carbon Nanotubes (CNTs) have garnered significant attention and have been extensively studied because of their unique electric, optical, thermodynamic, and mechanical properties. They are cylindrical structures formed by a single layer of carbon atoms that are rolled into a tube shape, usually with a diameter ranging from tens of nanometers. They may be classified into two main categories: Single-walled carbon nanotubes (SWCNTs) and Multi-walled carbon nanotubes (MWCNTs). Multi-walled carbon nanotubes (MWCNTs) consist of several concentric layers of rolled graphite, as seen in Figure 7.1. Single-walled carbon nanotubes (SWCNTs) have a high aspect ratio. In recent years, many studies have been conducted on the utilization of CNTs for administering therapies in both laboratory and real organisms (Poater et al., 2012).

CNTs possess a high aspect ratio, which distinguishes them from other spherical nanoparticles. This is due to their needle-like shape, which enables them to carry a huge amount of payloads down the length of the tubes without compromising their capacity to penetrate cells (Zhang et al., 2014). The CNTs have sufficient loading capacity to transport many therapies, such as medicines, genes, and molecule targeting, into a single cell to produce multiple effects. On the contrary, due to their exceptionally large surface area, robust mechanical qualities, and electrical conductivity, carbon nanotubes (CNTs) are highly suitable for use as nano-scaffolds and in the creation of three-dimensional nanocomposites (Son et al., 2007).

Recently, devices based on carbon nanotubes (CNTs) have been effectively employed in tissues engineering and therapeutic applications involving stem cells. These applications include cardiac treatment, bone production, muscle regeneration, and brain regeneration. In addition, due to the unique optical characteristics of carbon nanotubes (CNTs), such as their substantial absorption in the near-infrared spectrum, photoluminescence, and pronounced Raman shift (Tan et al., 2011). Carbon nanotubes (CNTs) are highly effective tools for detecting and photographing biological entities. Nano-devices that are targeted and based on carbon nanotubes (CNTs) have been created for selective imaging and sensing, thanks to the combination of CNTs' large surface area for binding molecular recognition molecules. Carbon nanotubes (CNTs) have a wide range of applications in several fields. This chapter outlines the techni□ues for preparing carbon nanotubes (CNTs) for their application in the field of medicine (Vashist et al., 2018).

Figure 7.1. *Schematics of a SWCNT and MWCNT.*

Source: Shivani Soni, Creative Commons License.

7.1. PREPARATION OF CNTS FOR USE IN MEDICINE

Learning Objectives

- Comprehend the process of covalent functionalization of CNTs
- Understand the process of non-covalent functionalization of CNTs for use in medicine

Carbon nanotubes (CNTs) that have not been treated demonstrate a high level of resistance to water as a result of the intrinsic metallic characteristics that they possess. For this reason, the process of modifying the surface of carbon nanotubes, which is also referred to as CNT functionalization, is essential for the usage of these nanotubes in medical applications. This is done to ensure that they are distributed equally in water-based solutions. The CNT modification techni□ues include both non-covalent and covalent approaches to the modification process. To apply a layer of amphiphilic chemicals, the non-covalent modification takes use of the hydrophobic features of carbon nanotubes (CNTs), notably the π-π interactions. The development of chemical bonds between carbon atoms on the surface of carbon nanotubes (CNTs) is an example of covalent modification (Masotti & Caporali, 2013).

7.1.1. Non-Covalent Functionalization

Functionalization of carbon nanotubes (CNTs) via non-covalent interactions may be done without altering the inherent sp^2-hybridized configuration of the nanotube sidewall. This allows for the preservation of the original electronic configuration and characteristics of CNTs. Various forms of noncovalent functionalization are currently being investigated (Habibizadeh et al., 2017).

7.1.1.1. π-π Stacking Interactions

Conjugated molecules can be non-covalently functionalized onto the graphitic wall of CNTs by π–π stacking interactions. Compounds containing the pyrene group, such as N-succinimidyl-1-pyrenebutanoate (Figure 7.2), can be strongly attached to the surface of single-walled carbon nanotubes (SWCNTs) by a kind of contact called π–π interaction (Yang et al., 2018). In addition, by using the succinimidyl ester group onto this bifunctional anchor, other molecules containing primary or secondary amines (such as ferritin, streptavidin, or biotin-polyethylene oxide-amine) were chemically bonded to the activated pyrene by a nucleophilic reaction with the amino group. By employing a comparable approach, other biomolecules such as proteins and DNA, as well as gold nanoparticles, were also attached to carbon nanotubes

(CNTs) (Chen et al., 2018), Phthalocyanines, porphyrins, and their derivatives can be immobilized on the surface of carbon nanotubes (CNTs) by π-π stacking interactions. This is seen in Figure 7.2. Evidence was provided to show that this particular composite material is capable of exhibiting photo-induced transfer of charge from the dye molecules to carbon nanotubes (CNTs). Furthermore, the researchers also identified a specific and deliberate interplay among the porphyrin derivative and CNTs (Zuo et al., 2013). Raman spectroscopy, near-infrared (NIR) absorption, and buck conductivity have demonstrated the preferential adsorption of porphyrin derivatives onto semiconductor nanotubes in a solubilized sample. This approach provides a practical means of distinguishing between semiconductive carbon nanotubes (CNTs) and metallic CNTs. Polymers possessing conjugated structures can also be connected to carbon nanotubes (CNTs) by π-π interactions (Figure 7.2). The robust connection between the nanotube and the conjugated polymer significantly enhanced the dispersibility of CNTs. Conjugated polymers are being utilized to effectively separate carbon nanotubes (CNTs) from their major contaminant, which is composed of amorphous graphitic shells. Amorphous carbon impurities have a tendency to settle out of the solution, while the nanotubes remain suspended (Adeli et al., 2013).

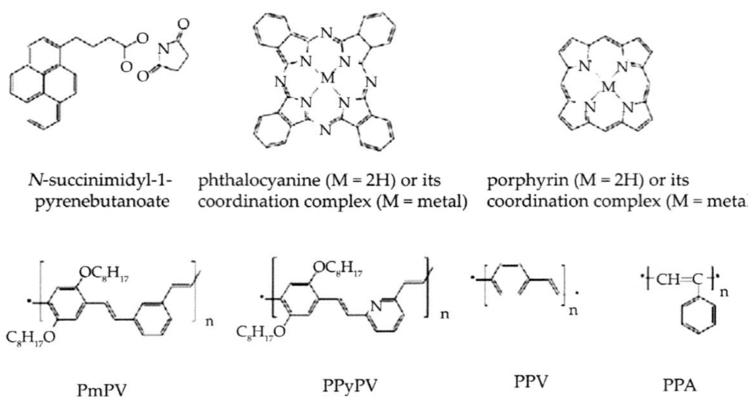

Figure 7.2. *Illustration of typical molecules that can non-covalently modify CNTS via π–π stacking interaction.*

Source: Shashi Bala, Creative Commons License.

7.1.1.2. Hydrophobic Interactions

Using hydrophobic interactions in water-based solutions, carbon nanotubes (CNTs) can be bound to amphiphilic chemicals, which is yet another method for non-covalently altering carbon nanotubes (CNTs). Within the context of this system, the amphiphilic molecules engage in non-covalent interactions with the aromatic surface of carbon nanotubes (CNTs) by utilizing their hydrophobic regions, while at the same time exposing their hydrophilic parts to the aqueous solution. This configuration serves the purpose of lowering the hydrophobic barrier that exists between the nanotubes and the polar materials that surround them (He et al., 2013). Hydrophobic interactions made it easier for single-walled carbon nanotubes (SWCNTs) to dissolve in water. This was accomplished by helical wrapping water-soluble polymers, specifically poly styrene sulfonate (PSS) and poly vinyl pyrrolidone (PVP), around the surface of the SWCNTs. The interaction between the polymer and the SWCNTs is reversible, and it is amenable to modification through the modification of the solvent solution. This connection is extremely stable and uniform across the whole surface of the nanotubes, and it does not depend on the quantity of extra polymer that is present in the solution. In addition, surfactants were affixed to carbon nanotubes (CNTs) by the use of hydrophobic interactions. Up until this point, numerous forms of surfactants, including anionic, cationic, and nonionic surfactants, have been utilized in order to disperse carbon nanotubes in a solution that contains water. Among them, it has been found that bile salt surfactants, more notably deoxycholic acid and taurodeoxycholic acid, do very well when it comes to dissolving single-walled carbon nanotubes (SWCNTs). The coulomb repulsive force that exists between surfactant-coated SWCNTs was responsible for the inhibition of aggregation within the material (Shao et al., 2013).

Furthermore, the presence of aromatic moieties in the hydrophobic portion of the amphiphilic molecule would result in extremely powerful interactions between the molecule and the carbon nanotubes (CNTs). As a result of this particular sort of aggregation, additional π-π interactions are generated between the aromatic component of the molecule that is being considered and the graphitic sidewalls of carbon nanotubes (CNTs). Additionally, the hydrophilic groups are exposed to the aqueous solution that surrounds the molecule. For instance, sodium dodecyl benzene sulfonate has demonstrated an improved ability to disperse carbon nanotubes (CNTs) in water due to its higher capacity. It was revealed that surfactants such as SDS, SDBS, DOC, or TDOC are quite effective at dispersing single-walled carbon nanotubes (SWCNTs) separately (Ijaz et al., 2023).

Additional attachments of biomolecules were made to carbon nanotubes by the use of hydrophobic interactions. To be more specific, a peptide that possessed amphiphilic characteristics was encased around the surface of a single-walled carbon nanotube (SWCNT). The peptide exhibited the ability to assume an α-helical structure when it was in an a□ueous solution. This resulted in the hydrophobic side of the helix being exposed to the surface of the nanotube. Using various techniques it was demonstrated that the peptide was responsible for the bundling of single-walled carbon nanotubes (SWCNTs). The nanotubes, which were wrapped in peptides, were arranged into a large-scale structure through the use of electrostatic interactions between neighboring nanotubes that were coated in peptides. The manipulation of the size and shape of the super-molecular structure is made possible by the control of the parameters that govern the interactions between peptides interacting with one another. In addition, it was demonstrated that the power to scatter individual SWCNTs was improved when the number of aromatic residues contained within the peptide sequence was doubled (Dubey et al., 2021).

Did You Know?

Carbon nanotubes (CNTs) are around 100 times stronger than steel but only a fraction of the weight, making them one of the strongest and lightest materials known.

7.1.2. Covalent Functionalization

CNTs include inherent properties, including little curvature, limited solubility, and a natural inclination to form bundles, which render this carbon nanomaterial resistant to chemical treatment. Several methods for covalently modifying carbon nanotubes (CNTs) have been developed. The methodologies for covalently functionalizing carbon nanotubes (CNTs) may be categorized into two primary groups based on the position of the functional groups:

i. Defect functionalization;

ii. Sidewall functionalization (Merum et al., 2017).

7.1.2.1. Defect Functionalization

Carbon nanotubes (CNTs) produced using any accessible technique invariably contain imperfections. The intrinsic defects of the sidewall consist of Stone-Wales defects, which are five- or seven-membered rings, as well as sp^3-hybridized defects and vacancies in the sp^2-hybridized six-membered ring carbon structure. CNTs can only tolerate a few flaws without compromising their macroscopic electrical and mechanical characteristics. Furthermore, the tips of the tubes, which take on a hemispherical fullerene shape, exhibit greater reactivity compared to the sidewalls due to their increased curvature (Wu et al., 2010). When exposed to powerful oxidizing chemicals such as $KMnO_4$ /H_2SO_4, O_2, $K_2Cr_2O_7$ /H_2SO_4, or OsO_4, carbon nanotubes (CNTs) can be fragmented and transformed into shorter tubes. Meanwhile, imperfections were introduced in the vicinity of the carbon nanotube (CNT) surface. Furthermore, the oxidization process resulted in the formation of oxygenated functional groups, including carboxylic acid, ketone, alcohol, and ester groups, at the ends and defect sites of the nanotubes. These functional groups played a crucial role in enhancing the functionality of carbon nanotubes (CNTs). Sonication may also induce flaws on the sidewall of carbon nanotubes (CNTs), which are very reactive and prone to additional chemical reactions (Jain et al., 2011).

SWCNTs were initially produced via carboxylic amidation using octadecylamine. Subsequently, lipophilic and hydrophilic Dendrons have been connected to CNTs via amide or ester bonds, resulting in a notable enhancement in the solubility of CNTs in organic or aqueous solutions. These groups can be later eliminated using either basic or acidic hydrolysis conditions. Furthermore, as a result of this functionalization process, the CNT bundles underwent significant exfoliation, resulting in the formation of individual nanotubes. This enabled additional modification and examination of the spectroscopic characteristics of individual single-walled carbon nanotubes (SWCNTs) (Eatemadi et al., 2014).

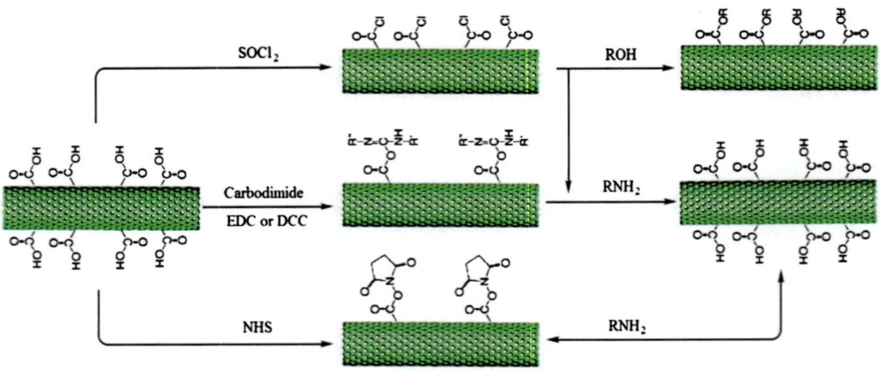

Figure 7.3. *Illustration of typical CNTs carboxylic acid derivatization routes.*

Source: Amandeep Salhotra, Creative Commons License.

7.1.2.2. Sidewall Functionalization

The sidewalls of carbon nanotubes (CNTs) contain a structure made up of six-membered rings with an aromatic nature. So far, numerous sidewall reactions have been developed, such as radical addition, fluorination, nucleophilic and electrophilic addition, and cycloaddition. Non-planar conjugated molecules exhibit reactivity that is primarily affected by two key factors: the pyramidalization of the attached carbon atom and the misalignment of π-orbitals between neighboring pairs of conjugate carbon atoms. Similar to a chemist's analysis, the strain in the sidewalls of CNTs is primarily a result of the misalignment of π-orbitals, in contrast to fullerene where strain is mainly caused by pyramidalization. Smaller nanotubes showcase heightened chemical reactivity, due to the intriguing connection between π-orbital misalignment and pyramidalization in relation to nanotube diameter. Chemical reactions occurring on the sidewalls of carbon nanotubes (CNTs) can have a detrimental effect on their conjugated structure. This disruption can be observed and analyzed using techniques such as Raman spectroscopy or absorption spectroscopy (Khabashesku et al., 2005).

The Raman spectra of virgin single-walled carbon nanotubes (SWCNTs) exhibit three distinct characteristic modes:

i. The radial breathing mode (ω_r) below 350 cm^{-1};
ii. The tangential mode (ω_t) around 1,550–1,600 cm^{-1}; and
iii. The disorder mode (sp^3) at 1,250–1,450 cm^{-1} (Hamilton et al., 2013).

The radial mode demonstrated a strong correlation with the diameter of the nanotube. The disorder mode's intensity showed a direct correlation with the number of carbon atoms that were sp^3-hybridized in the single-walled carbon nanotubes (SWCNTs). This information offers valuable insights into the extent of functionalization on the sidewalls, where the carbon atoms that were originally bonded to functional groups underwent a transformation from sp^2-hybridized to sp^3-hybridized. The absorption spectra revealed changes in the sidewall of single-walled carbon nanotubes (SWCNTs) that were caused by covalent modifications. The reaction caused the conjugated structure to break, resulting in the lack of optical transitions between van Hove singularities in the electronic density of state (DOS) of carbon nanotubes (CNTs) (Peng et al., 2003).

7.1.2.3. Fluorination

Fluorinating single-walled carbon nanotubes (SWCNTs) requires the interaction of SWCNTs with elemental fluorine at temperatures ranging from 150 °C to 600 °C. Through careful control of the reaction parameters, such as time and temperature, one can alter the stoichiometry. Through careful analysis of absorption and the Raman spectrum, the electrical disturbance of fluorinated single-walled carbon nanotubes (SWCNTs) was revealed. Through the process of fluorination, the bundling of single-walled carbon nanotubes (SWCNTs) was effectively disrupted, resulting in the dispersion of individual nanotubes (Salah et al., 2021). This treatment also greatly improved the solubility of SWCNTs in various solvents such as dimethylformamide (DMF), tetrahydrofuran (THF), and alcohols. Through careful experimentation, the interaction between hydrazine and the C-F bonds had a remarkable effect on the conductance and spectroscopic properties of the single-walled carbon nanotubes (SWCNTs), leading to their almost complete restoration. In the study, it was observed that the Fermi energy level and conduction band energy of fluorinated SWCNTs were significantly lower than those of the original SWCNTs. Fluorinated CNTs were discovered to possess exceptional electron accepting properties in comparison to virgin nanotubes. Consequently, they showcased heightened chemical reactivity when exposed to powerful nucleophilic reagents like Grignard, alkyllithium reagents, and metal alkoxides, as illustrated in Figure 7.4 (Song et al., 2015) (Figure 7.5).

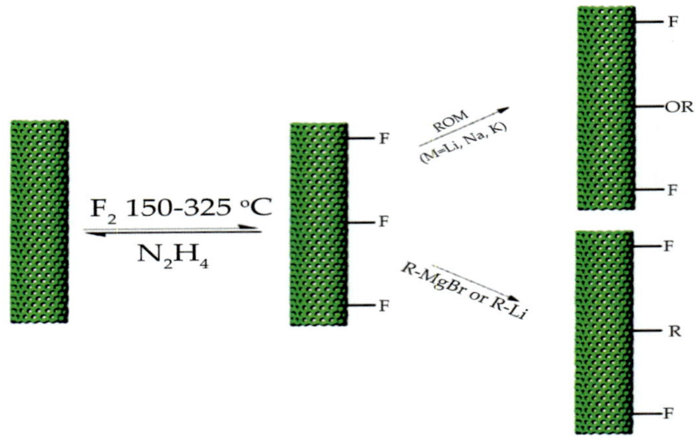

Figure 7.4. *Fluorination of CNTs and further modification routes.*

Source: Mrutyunjay Suar, Creative Commons License.

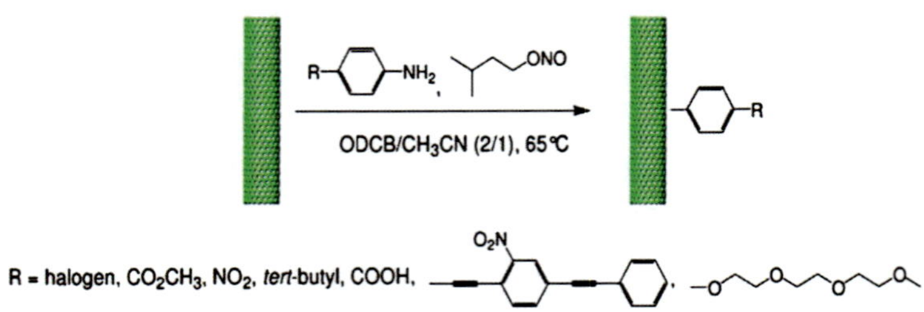

Figure 7.5. *Example of strategy to functionalize SWCNTs with diazomiun compounds.*

Source: Shashi Bala, Creative Commons License.

7.1.2.4. Nucleophile and Electrophilic Addition

Due to the curved shape of the sidewall, carbon nanotubes (CNTs) exhibit a significant affinity for electrons. These substances have the ability to interact with potent nucleophile reagents like sec-butyl lithium, nucleophile carbine, and polymeric carbanions. These reactions result in the creation of different functionalized CNTs, as illustrated in Figure 7.6. In addition, the successful electrophilic addition occurred through the interaction between chloroform and the sidewall of carbon nanotubes (CNTs) in the presence of a Lewis acid (Sheikhi et al., 2018).

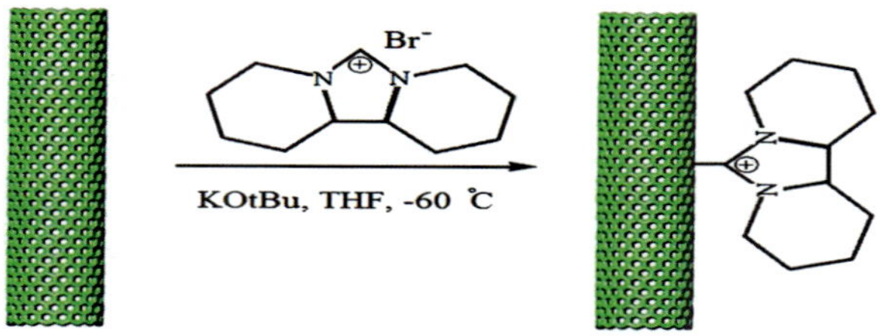

Figure 7.6. Reaction of SWCNTs with nucleophilic carbene.

Source: Raj Kishore, Creative Commons License.

Remember

CNTs can conduct electricity better than copper, making them potential candidates for ultra-miniature electrical circuits in future electronics and medical devices.

7.2. APPLICATIONS OF CNTS IN MEDICINE

Learning Objectives

- Comprehend the mechanism by which CNTs facilitate cell penetration and their potential benefits in medical treatments
- Understand how CNTs are used in drug delivery systems
- Comprehend the antioxidant properties of CNTs

There are several envisioned uses of carbon nanotubes (CNTs) in healthcare and related domains, and many research organizations are now engaged in these endeavors. A significant catalyst for biomedical research is the capacity of carbon nanotubes (CNTs) to infiltrate cells. It is important to emphasize that seemingly conflicting outcomes might be documented, mostly due to the significant variations in structure among the materials being investigated. Specific macroscopic distinctions, such as the use of SWCNTs or MWCNTs, the sort of surface alteration, and the level of aggregation, must be documented. However, the diameter and length figures are frequently overlooked and, when they are included, they typically indicate an approximate average. The subsequent sections describe notable instances of contemporary biological knowledge developed based on their respective fields of application, without considering the specific types of carbon nanotube (CNT) materials employed (Sireesha et al., 2018).

7.2.1. Cell Penetration

It is widely acknowledged that carbon nanotubes (CNTs) may readily go into several types of cells. The potential processes involved include endocytosis, phagocytosis, or membrane piercing (Amenta & Aschberger, 2015).

In all research conducted, the outcomes were quite promising; indeed, the carbon nanotubes infiltrate cells without causing any deleterious effects. The cells maintained their structural integrity, fluidity of the cytoplasm, and rate of cell division. Notably, the BY-2 cells did not contain the single-stranded DNA when supplied alone. The cellular distribution varies: the Ox-SWCNT/FITC is localized in the vacuoles, whilst the SWCNT/ssDNA can be observed in the cytoplasmic threads. The internalization process was shown to be reliant on both time and temperature, indicating the potential participation of endocytosis. It is worth noting that in plant cells, a reduction in temperature can inhibit endocytosis (Cheung et al., 2010).

7.2.2. DRUG DELIVERY

Additionally, as was mentioned earlier, the capability of filling the interior cavity of carbon nanotubes (CNTs) presents a fresh approach to the process of building drug reservoirs. Using carboplatin as a medicinal drug, Hampel et al. (2008) conducted research in this particular field of inquiry. Oxidized multi-walled carbon nanotubes (MWCNTs) of varied lengths and diameters were utilized by the researchers in their study. For the purpose of starting the process of opening the tubes, the oxidation operation was carried out. There is a strong indication that the incorporation is dependent on the temperature. For the purpose of confirming that the medication was retained within CNTs, the researchers applied a number of different techniques, one of which was X-ray photoelectron spectroscopy. However, they also found trace levels of carboplatin that had been deposited outside of the body (Rahamathulla et al., 2021) (Figure 7.7).

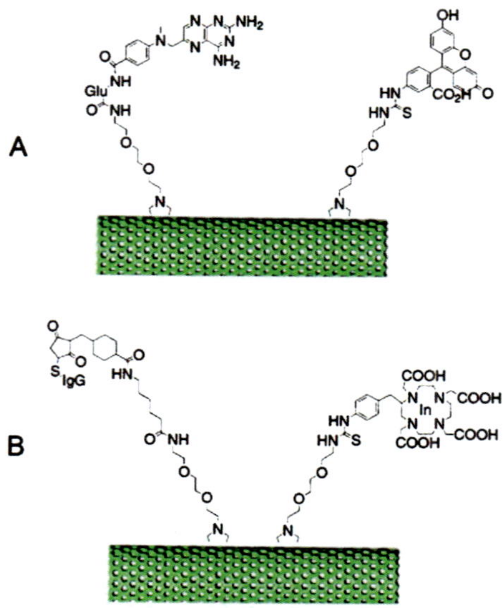

Figure 7.7. *Molecular structures of bifunctional CNT derivatives.*

Source: Neeraj Vij, Creative Commons License.

Single-walled carbon nanotubes (SWCNTs) that have been covered with modified phospholipids (PL-PEG-NH_2) can connect a platinum (IV) compound to one of the axial Pt ligands through the amine residue. The compound has minimal toxicity against testicular cancer cells on its own but demonstrates a substantial cytotoxic effect when it is bound to the surface of amine-functionalized soluble SWCNTs. The inclusion of CNTs facilitates the cellular absorption of the prodrug, leading to its confinement within endosomes. These endosomes have a lower pH compared to the cell incubation media. The acidic conditions promote the conversion of Pt (IV) to Pt (II) and the liberation of active cisplatin, resulting in the detachment of axial ligands (shown in Figure 7.8) (Elhissi et al., 2012).

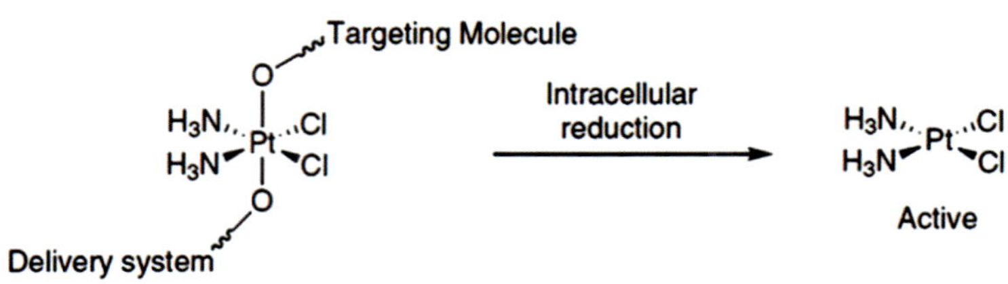

Figure 7.8. *Schematic of cisplatin release mechanism.*

Source: Shivani Soni, Creative Commons License.

In a more recent study, the authors introduced a novel complex where a platinum axial ligand is connected to folic acid, serving as a targeting component (Figure 7.9, A). Additionally, the formulation of large-molecule complexes is expected to enhance the duration of blood circulation. The researchers investigated human choriocarcinoma and human nasopharyngeal carcinoma cell lines that express the Folate Receptor (FR). They used testicular carcinoma cells (NTera-2) that do not express the Folate Receptor as a negative control. The results showed that the FR-positive cells were more significantly impacted. Furthermore, the prodrug that was produced showed much higher activity compared to cisplatin alone. It formed cisplatin 1, 2-d (GpG) intrastrand cross-links to nuclear DNA after undergoing intracellular reduction. Recent studies have shown that the adsorption of doxorubicin on the surface of multi-walled carbon nanotubes (MWCNT) through π–π interaction (as shown in Figure 7.9, B) enhances its effectiveness compared to the free form. This is likely due to improved transport into cells (Rosen & Elman, 2009).

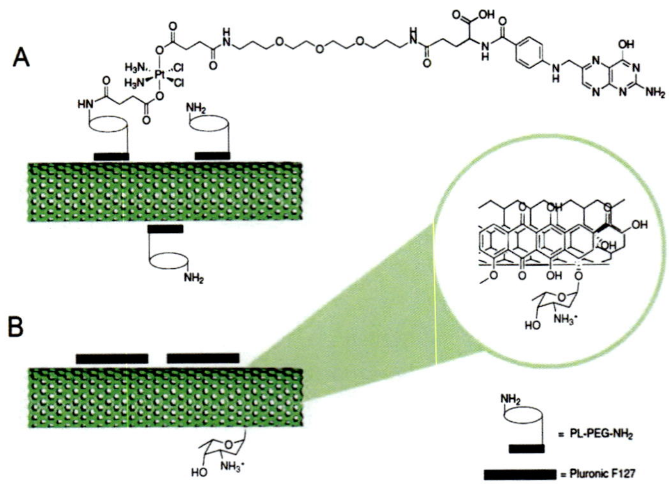

Figure 7.9. *Non-covalent attachments of anti-tumoral drugs to CNTs.*

Source: Amandeep Salhotra, Creative Commons License.

7.2.3. Gene Delivery

In the process of developing one of the first techniques for conveying genetic material, Lu et al. (2003) formed complexes between RNA polymer poly (rU) and SWCNTs through random interactions. This was one of the first approaches that was established. As a consequence of the translocation that took place in MCF7 breast cancer cells, genetic material was found to be present in both the cellular membrane and the nuclear membrane. The primary objective of the research was to investigate the potential of using covalent derivative single-walled carbon nanotubes (SWCNTs) as a delivery system for telomerase reverse transcriptase small interfering RNA (TERT-siRNA) into cells. Through the process of amidation, the carboxylic groups on the nanotubes were converted into SWCNTs. This was accomplished by employing a chain that contained a terminal ammonium group. When determining the impact, both the proliferation and growth of tumor cells in laboratory conditions and in living creatures were taken into consideration. The successful inhibition of the particular gene expression serves as evidence of the efficient intracellular transport of siRNA, which is expelled from nanotubes in a localized manner and has the effect that was intended for it to have (Zare et al., 2021).

Human T cells and peripheral blood mononuclear cells were both subjected to the administration of siRNA by Dai and his colleagues working together. This particular occurrence involved the use of PL-PEG-NH_2 to cover SWCNTs for the second time. It was found that the efficiency of the PL-PEG construct varied depending on the functionalization and hydrophilicity, both of which were affected by the length of the PEG chain that was introduced. The use of short SWCNTs that had been modified with poly-l-polyethylene glycol (PL-PEG) using non-covalent techniques was implemented. For this particular case, a disulfide bond was utilized to establish a cleavable connection between the PLPEG unit and the siRNA that was being delivered. It was discovered that the functioning of siRNA that was achieved in this model was more powerful than that of lipofectamine, which was used as the control transfection agent (Karimi et al., 2015).

SWCNTs that have been oxidized and have positive charges have the potential to interact with fluorescein-labeled double-stranded DNA (dsDNA-FAM). As an additional step, the SWCNTs were coated with phospholipids that had been modified with folic acid (PL-PEG-NH2) by wrapping the alkyl chains around them (Figure 7.10). The tumor cells, which contained an excessive amount of folate receptors, took up the multifunctional system in a selective manner. Fluorescence imaging provided evidence that the oligonucleotide was delivered into the cells as a consequence of this process (Bates & Kostarelos, 2013).

To transport single-stranded DNA (ssDNA) probes into cells, SWCNTs were utilized as carriers. As a consequence of this, the oligonucleotides were afforded a higher level of protection against the degradation inflicted by nucleases. Despite the fact that it is protected from enzymatic breakage and disruption induced by nucleic acid binding proteins, single-stranded DNA (ssDNA) continues to successfully carry out its function of targeting messenger

RNA (mRNA). Kateb et al. (2007) had earlier claimed that carbon nanotubes (CNTs) may be used as carriers for oligonucleotides (Mohajeri et al., 2019).

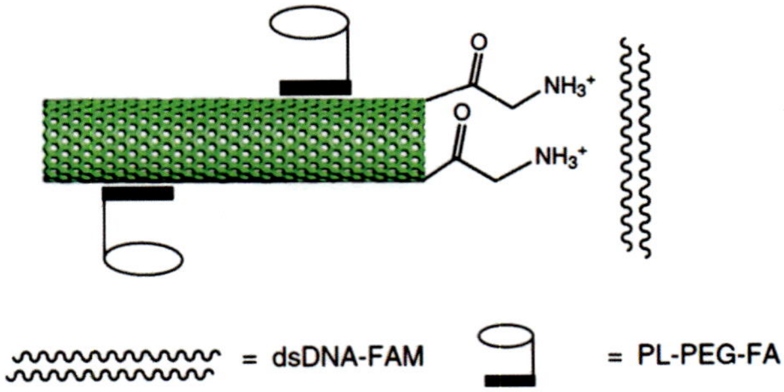

Figure 7.10. Illustration of CNT-based multifunctional DNA delivered system.

Source: Om Singh, Creative Commons License.

Practice Problem

Describe the mechanism by which CNTs can enhance gene delivery into cells. What are the advantages of using CNTs for this purpose?

Solution to Practice Problem

Mechanism:

i. Complex Formation: CNTs can form complexes with nucleic acids (DNA or RNA) through non-covalent interactions such as π-π stacking, electrostatic interactions, or hydrogen bonding.

ii. Cell Penetration: The functionalized CNT-nucleic acid complex can penetrate the cell membrane via endocytosis or direct membrane translocation

iii. Release: Once inside the cell, the nucleic acids are released from the CNTs either through a change in the cellular environment (pH, enzymes) or by the breakdown of the CNT complex

Advantages:

i. Efficient Delivery: CNTs can efficiently deliver genetic material into cells due to their high aspect ratio and ability to cross cell membranes

ii. Protection of Genetic Material: CNTs can protect nucleic acids from degradation by nucleases during delivery

iii. Targeted Delivery: Functionalization allows for the attachment of targeting ligands, ensuring that the genetic material is delivered specifically to the target cells or tissues

iv. Versatility: CNTs can be used to deliver various types of genetic material, including plasmid DNA, siRNA, and mRNA

7.2.4. Antioxidant Properties of CNTs

When carbon nanotubes (CNTs) were introduced into the lungs, they caused an inflammatory and fibrotic reaction. This reaction was believed to be caused by oxidative stress resulting from free radicals, however, there was no actual proof of reactive oxygen species (ROS) formation from CNTs. In addition, Fenoglio, Tomatis, and Fubini (2006) found that MWCNTs do not produce oxygen-based or carbon-based free radicals when exposed to H_2O_2 or water. However, when radicals are introduced, CNTs function as scavengers, removing the radicals. Tour and his colleagues have recently published on the antioxidant capabilities of CNTs (Fu et al., 2017).

SWCNTs and US-SWCNTs were modified using butyrate hydroxyl toluene (BHT) utilizing two distinct methods: triazene covalent attachment to the sidewalls of pluronic-wrapped SWCNTs, or amidation of carboxylic residues for US-SWCNT derivatives. The oxygen radical scavenging capacity of the various compounds has been assessed, yielding intriguing findings, despite the varying reactions shown by the compounds. In the first scenario, the pluronic-coated SWCNTs demonstrated more efficiency compared to the e□uivalent derivatives of butylated hydroxytoluene (BHT). However, in the case of oxidized ultrasonicated SWCNTs (US-SWCNTs), the antioxidant activity improved with increased levels of BHT residues. This recent discovery opens up opportunities for the use of SWCNT as an innovative medicinal treatment in the realm of antioxidants. Nevertheless, it is important to note that a recent investigation has presented conflicting findings about the toxicity of MWCNTs in A549 cells. The article suggests that this toxicity can be mitigated by administering antioxidants before exposure, which reduces the generation of reactive oxygen species (ROS) and the expression of the interleukin-8 gene (Rizzo et al., 2019).

7.3. TOXICITY OF CARBON NANOTUBES

Learning Objectives

- Understand the potential toxicity issues associated with CNTs when used in medical applications

A significant discussion and several concerns over the toxicity of carbon nanotubes (CNTs) emerged several years ago. However, the available knowledge remains incomplete and seems inconsistent. It is challenging to categorize definite conclusions because the research conducted encompasses a broad range of derivatives with varying properties. Generally, it can be stated that toxicity diminishes in proportion to the rise in dispersion along with solubility level. However, additional exploration is required in this domain (Francis & Devasena, 2018).

Investigations were conducted to study the hydroxyl radical scavenging processes of modified carbon nanotubes (CNTs) using electron paramagnetic resonance (EPR). Through careful analysis, it was found that the untreated CNTs exhibited activity against hydroxyl radicals, a characteristic that was no longer present in the purified and annealed sample. Nevertheless, the metal-deprived carbon structure regained its activity due to the presence of induced structural defects. Experiments were carried out on rat lung epithelial cells in a controlled laboratory environment to assess the potential genetic harm caused by different carbon nanotube (CNT) preparations. Through a careful heating process, the acute pulmonary toxicity and genotoxicity of CNTs were significantly reduced. However, it is worth noting that these beneficial effects were reversed when the CNTs underwent grinding. It appears that the presence of defective sites within the carbon framework is the primary reason behind the inherent toxicity of CNTs. The scavenging activity is connected to the presence of imperfections and seems to be related to the genotoxic and inflammatory properties of CNT (Kayat et al., 2011).

In their study, Tabet, Bussy, and Lanone (2008) utilized a combination of multi-walled carbon nanotubes (MWCNT) in ethanol, dipalmitoyl lecithin, and phosphate-buffered saline (PBS) to examine the potential negative impacts of CNTs on the A549 human epithelial cell line. PBS promotes the formation of clusters on the cell surface. Nevertheless, in every case, MWCNTs decrease cellular metabolism without inducing any damage to cellular membranes or triggering cell death. However, asbestos fibers infiltrate the cells and stimulate an increase in apoptosis. In a study conducted by Wick et al. (2007), the impact of carbon nanotubes (CNTs) on human MSTO-211H cells was investigated. The researchers utilized CNTs that exhibited different levels of aggregation to assess their cytotoxicity. Well-dispersed CNTs, achieved through the use of polyoxyethylene sorbitan monooleate, demonstrated reduced toxicity in comparison to asbestos. In addition, clusters of CNTs that resembled ropes had a stronger impact on cell toxicity compared to individual asbestos fibers (Muller et al., 2006).

Guo, Zhang, & Li (2008) carried out a study where they made some changes to MWCNTs by utilizing glucosamine and decylamine with the help of c-ray irradiation. After obtaining the derivatives, namely g-MWCNT and d-MWCNT, a cytotoxicity test was performed on Tetrahymena pyriformis. The growth inhibition of the decylamine derivative is influenced by the dosage, as it is affected by the toxicity of the amine. The toxicity is enabled by the nanotubes, which facilitate the entry of the amine into the cells. The comparison between pure multi-walled carbon nanotubes (MWCNTs) and graphitized MWCNTs (g-MWCNTs) revealed that the latter promoted growth.

It seems that the hydrophilic molecule can bind with peptone, a substance present in the medium, and aid in its transportation into cells. It is important to consider the potential interactions between carbon nanotubes (CNTs) and the various components of the culture media when evaluating the cytotoxicity of CNTs. Researchers have also directed their attention towards studying the ciliated protozoan Tetrahymena thermophile, due to the limited knowledge of the effects of CNTs on natural ecosystems. This organism has the remarkable ability to control microbial populations by consuming and digesting bacteria. This organism plays a crucial role in the treatment of wastewater and serves as a dependable indicator of sewage effluent □uality. Microorganisms have been found to contain single-walled carbon nanotubes (SWCNTs), which lead to the formation of clusters and the loss of their ability to communicate with bacteria. The mutagenicity assay involved the utilization of Salmonella typhimurium and Escherichia coli strains, similar to a chemist conducting experiments. No evidence of mutagenic activity was observed, even when metabolic activators were present (Chetyrkina et al., 2022).

Several experiments have been carried out to assess the genotoxicity of MWCNTs, including studies on rats and lung epithelial cells. After three days of administering carbon nanotubes directly into the trachea, there is an observed increase in micronuclei in type II pneumocytes, which is dependent on the dosage. Results from the in vitro tests showed that MWCNTs can cause clastogenic and aneugenic effects. In experiments, it was observed that MWCNTs have the ability to increase the occurrence of genetic mutations caused by DNA breaks, as well as the alteration of entire chromosomes. Reactive oxygen species (ROS) are often responsible for the clastogenic effect, which can be linked to the presence of metallic nanoparticles like iron (Fe) or cobalt (Co). Nevertheless, the potential antioxidant properties of multi-walled carbon nanotubes (MWCNT) indicate the existence of alternative mechanisms, underscoring the importance of further research in these systems (Kolosnjaj et al., 2007).

SUMMARY

- Carbon nanotubes (CNTs) have demonstrated a wide range of physical, chemical, and mechanical qualities that make them appropriate for many applications. Over the past decade, there has been significant advancement in the use of carbon nanotubes (CNTs) in medicinal applications.
- This chapter has provided an overview of the chemical and physical techniques utilized to manufacture carbon nanotubes (CNTs) for medical applications. These approaches include attaching targeting molecules to carbon nanotubes (CNTs) to provide tailored drug administration, selective imaging, and other therapeutic applications.
- A high volume of research has been published so far on the toxicity of CNTs. Nevertheless, the reported data exhibit inconsistencies. The variability in dispersion status, size, length of tubes, metal impurities, and functionalization procedures of carbon nanotubes (CNTs) employed in these investigations is the cause.
- Furthermore, the employment of various analysis methodologies in the assessment of toxicity research on carbon nanotubes (CNTs) also contributes to discrepancies. Although there are differences, there is a consensus that well-distributed carbon nanotubes (CNTs) have minimal or no toxicity when tested in laboratory settings and in living organisms, making them valuable for medicinal purposes.
- There is a pressing re□uirement for conducting long-term investigations on the absorption, deposition, metabolism, and excretion (ADME) of CNTs. CNT-based treatments can only be implemented therapeutically if the ambiguity around CNT toxicity is resolved.

MULTIPLE CHOICE QUESTIONS

1. Which method involves chemically modifying the surface of CNTs to attach functional groups?

a. Non-covalent functionalization

b. Covalent functionalization

c. Physical adsorption

d. Electrostatic interaction

2. Non-covalent functionalization of CNTs relies primarily on:

a. Chemical reactions

b. Physical interactions
c. Ionic bonds
d. Metal coordination

3. Which application involves using CNTs to facilitate the transport of therapeutic molecules across cellular membranes?

a. Gene delivery
b. Cell penetration
c. Imaging contrast agents
d. Tissue engineering

4. In the context of CNTs, drug delivery aims to:

a. Improve therapeutic efficacy and minimize side effects
b. Increase chemical reactivity of drugs
c. Enhance electrical properties of drugs
d. Strengthen the mechanical properties of tissues

5. Which of the following is a major concern regarding the use of CNTs in medicine?

a. High cost of production
b. Electrical conductivity
c. Toxicity and biocompatibility
d. Structural integrity

REVIEW QUESTIONS

1. What are carbon nanotubes (CNTs) and why are they significant in nanomedicine?
2. Explain the difference between covalent and non-covalent functionalization of CNTs.
3. What are some common molecules used in non-covalent functionalization of CNTs, and how do they interact with the CNT surface?
4. Discuss the role of CNTs in drug delivery systems. How do they improve therapeutic outcomes?
5. What are the main toxicity concerns associated with the use of CNTs in medicine? Provide examples of how these concerns are being addressed in research?

Answers to Multiple Questions

1. (b); 2. (b); 3. (b); 4. (a); 5. (c)

REFERENCES

1. Adeli, M., Soleyman, R., Beiranvand, Z., & Madani, F. (2013). Carbon nanotubes in cancer therapy: A more precise look at the role of carbon nanotube–polymer interactions. *Chemical Society Reviews*, *42*(12), 5231–5256.
2. Amenta, V., & Aschberger, K. (2015). Carbon nanotubes: Potential medical applications and safety concerns. *Wiley Interdisciplinary Reviews: Nanomedicine and Nanobiotechnology*, *7*(3), 371–386.
3. Bates, K., & Kostarelos, K. (2013). Carbon nanotubes as vectors for gene therapy: Past achievements, present challenges and future goals. *Advanced Drug Delivery Reviews*, *65*(15), 2023–2033.
4. Chen, T., Li, M., & Liu, J. (2018). π–π stacking interaction: A nondestructive and facile means in material engineering for bioapplications. *Crystal Growth & Design*, *18*(5), 2765–2783.
5. Chetyrkina, M. R., Fedorov, F. S., & Nasibulin, A. G. (2022). In vitro toxicity of carbon nanotubes: A systematic review. *RSC Advances*, *12*(25), 16235–16256.
6. Cheung, W., Pontoriero, F., Taratula, O., Chen, A. M., & He, H. (2010). DNA and carbon nanotubes as medicine. *Advanced Drug Delivery Reviews*, *62*(6), 633–649.
7. Dubey, R., Dutta, D., Sarkar, A., & Chattopadhyay, P. (2021). Functionalized carbon nanotubes: Synthesis, properties and applications in water purification, drug delivery, and material and biomedical sciences. *Nanoscale Advances*, *3*(20), 5722–5744.
8. Eatemadi, A., Daraee, H., Karimkhanloo, H., Kouhi, M., Zarghami, N., Akbarzadeh, A., & Joo, S. W. (2014). Carbon nanotubes: Properties, synthesis, purification, and medical applications. *Nanoscale Research Letters*, *9*, 1–13.
9. Elhissi, A., Ahmed, W., Dhanak, V., & Subramani, K. (2012). Carbon nanotubes in cancer therapy and drug delivery. In *Emerging Nanotechnologies in Dentistry* (Vol. 2, pp. 347–363). William Andrew Publishing.
10. Fenoglio, I., Tomatis, M., Lison, D., Muller, J., Fonseca, A., Nagy, J. B., & Fubini, B. (2006). Reactivity of carbon nanotubes: Free radical generation

or scavenging activity? *Free Radical Biology and Medicine*, *40*(7), 1227–1233.

11. Francis, A. P., & Devasena, T. (2018). Toxicity of carbon nanotubes: A review. *Toxicology and Industrial Health*, *34*(3), 200–210.
12. Fu, Y., Yang, C., Lvov, Y. M., Zhang, L., & Wang, W. (2017). Antioxidant sustained release from carbon nanotubes for preparation of highly aging resistant rubber. *Chemical Engineering Journal*, *328*, 536–545.
13. Guo, J., Zhang, X., Zhang, S., Zhu, Y., & Li, W. (2008). The different bio-effects of functionalized multi-walled carbon nanotubes on *Tetrahymena pyriformis*. *Current Nanoscience*, *4*(3), 240–245.
14. Habibizadeh, M., Rostamizadeh, K., Dalali, N., & Ramazani, A. (2017). Preparation and characterization of PEGylated multiwall carbon nanotubes as covalently conjugated and non-covalent drug carrier: A comparative study. *Materials Science and Engineering: C*, *74*, 1–9.
15. Hamilton Jr, R. F., Xiang, C., Li, M., Ka, I., Yang, F., Ma, D., & Holian, A. (2013). Purification and sidewall functionalization of multiwalled carbon nanotubes and resulting bioactivity in two macrophage models. *Inhalation Toxicology*, *25*(4), 199–210.
16. Hampel, S., Kunze, D., Haase, D., Krämer, K., Rauschenbach, M., Ritschel, M., & Büchner, B. (2008). Carbon nanotubes filled with a chemotherapeutic agent: A nanocarrier mediates inhibition of tumor cell growth. *Nanotechnology*, *5*(3), 654–656.
17. He, H., Pham-Huy, L. A., Dramou, P., Xiao, D., Zuo, P., & Pham-Huy, C. (2013). Carbon nanotubes: Applications in pharmacy and medicine. *BioMed Research International*, *2013*(1), 578290.
18. Ijaz, H., Mahmood, A., Abdel-Daim, M. M., Sarfraz, R. M., Zaman, M., Zafar, N., & Benguerba, Y. (2023). Review on carbon nanotubes (CNTs) and their chemical and physical characteristics, with particular emphasis on potential applications in biomedicine. *Inorganic Chemistry Communications*, *7*(2), 111020–111025.
19. Jain, S., Thakare, V. S., Das, M., Godugu, C., Jain, A. K., Mathur, R., & Mishra, A. K. (2011). Toxicity of multiwalled carbon nanotubes with end defects critically depends on their functionalization density. *Chemical Research in Toxicology*, *24*(11), 2028–2039.
20. Karimi, M., Solati, N., Ghasemi, A., Estiar, M. A., Hashemkhani, M., Kiani, P., & Hamblin, M. R. (2015). Carbon nanotube's part II: A remarkable carrier for drug and gene delivery. *Expert Opinion on Drug Delivery*, *12*(7), 1089–1105.

21. Kateb, B., Van Handel, M., Zhang, L., Bronikowski, M. J., Manohara, H., & Badie, B. (2007). Internalization of MWCNTs by microglia: Possible application in immunotherapy of brain tumors. *Neuroimage*, *37*, 9–17.
22. Kayat, J., Gajbhiye, V., Tekade, R. K., & Jain, N. K. (2011). Pulmonary toxicity of carbon nanotubes: A systematic report. *Nanomedicine: Nanotechnology, Biology and Medicine*, *7*(1), 40–49.
23. Khabashesku, V. N., Margrave, J. L., & Barrera, E. V. (2005). Functionalized carbon nanotubes and nanodiamonds for engineering and biomedical applications. *Diamond and Related Materials*, *14*(3–7), 859–866.
24. Kolosnjaj, J., Szwarc, H., & Moussa, F. (2007). Toxicity studies of carbon nanotubes. *Bio-Applications of Nanoparticles*, *7*, 181–204.
25. Lu, X., Tian, F., Xu, X., Wang, N., & Zhang, Q. (2003). A theoretical exploration of the 1, 3-dipolar cycloadditions onto the sidewalls of (n, n) armchair single-wall carbon nanotubes. *Journal of the American Chemical Society*, *125*(34), 10459–10464.
26. Masotti, A., & Caporali, A. (2013). Preparation of magnetic carbon nanotubes (Mag-CNTs) for biomedical and biotechnological applications. *International Journal of Molecular Sciences*, *14*(12), 24619–24642.
27. Merum, S., Veluru, J. B., & Seeram, R. (2017). Functionalized carbon nanotubes in bio-world: Applications, limitations and future directions. *Materials Science and Engineering: B*, *223*, 43–63.
28. Mohajeri, M., Behnam, B., & Sahebkar, A. (2019). Biomedical applications of carbon nanomaterials: Drug and gene delivery potentials. *Journal of Cellular Physiology*, *234*(1), 298–319.
29. Muller, J., Huaux, F., & Lison, D. (2006). Respiratory toxicity of carbon nanotubes: How worried should we be? *Carbon*, *44*(6), 1048–1056.
30. Peng, H., Alemany, L. B., Margrave, J. L., & Khabashesku, V. N. (2003). Sidewall carboxylic acid functionalization of single-walled carbon nanotubes. *Journal of the American Chemical Society*, *125*(49), 15174–15182.
31. Poater, A., Saliner, A. G., Cavallo, L., Poch, M., Solà, M., & Worth, A. P. (2012). Tuning the electronic properties by width and length modifications of narrow-diameter carbon nanotubes for nanomedicine. *Current Medicinal Chemistry*, *19*(30), 5219–5225.
32. Rahamathulla, M., Bhosale, R. R., Osmani, R. A., Mahima, K. C., Johnson, A. P., Hani, U., & Gangadharappa, H. V. (2021). Carbon nanotubes: Current perspectives on diverse applications in targeted drug delivery and therapies. *Materials*, *14*(21), 6707.

33. Rizzo, C., Marullo, S., Dintcheva, N. T., Gambarotti, C., Billeci, F., & D'Anna, F. (2019). Ionic liquid gels and antioxidant carbon nanotubes: Hybrid soft materials with improved radical scavenging activity. *Journal of Colloid and Interface Science*, *556*, 628–639.

34. Rosen, Y., & Elman, N. M. (2009). Carbon nanotubes in drug delivery: Focus on infectious diseases. *Expert Opinion on Drug Delivery*, *6*(5), 517–530.

35. Salah, L. S., Ouslimani, N., Bousba, D., Huynen, I., Danlée, Y., & Aksas, H. (2021). Carbon nanotubes (CNTs) from synthesis to functionalized (CNTs) using conventional and new chemical approaches. *Journal of Nanomaterials*, *2021*(1), 4972770.

36. Shao, W., Arghya, P., Yiyong, M., Rodes, L., & Prakash, S. (2013). Carbon nanotubes for use in medicine: Potentials and limitations. *Syntheses and Applications of Carbon Nanotubes and Their Composites*, *13*, 285–311.

37. Sheikhi, M., Shahab, S., Khaleghian, M., & Kumar, R. (2018). Interaction between new anti-cancer drug syndros and CNT (6, 6–6) nanotube for medical applications: Geometry optimization, molecular structure, spectroscopic (NMR, UV/Vis, excited state), FMO, MEP and HOMO-LUMO investigation. *Applied Surface Science*, *434*, 504–513.

38. Sireesha, M., Jagadeesh Babu, V., Kranthi Kiran, A. S., & Ramakrishna, S. (2018). A review on carbon nanotubes in biosensor devices and their applications in medicine. *Nanocomposites*, *4*(2), 36–57.

39. Son, S. J., Bai, X., & Lee, S. B. (2007). Inorganic hollow nanoparticles and nanotubes in nanomedicine: Part 2: Imaging, diagnostic, and therapeutic applications. *Drug Discovery Today*, *12*(15–16), 657–663.

40. Song, K., Gao, A., Cheng, X., & Xie, K. (2015). Preparation of the superhydrophobic nano-hybrid membrane containing carbon nanotube based on chitosan and its antibacterial activity. *Carbohydrate Polymers*, *130*, 381–387.

41. Tabet, L., Bussy, C., Amara, N., Setyan, A., Grodet, A., Rossi, M. J., & Lanone, S. (2008). Adverse effects of industrial multiwalled carbon nanotubes on human pulmonary cells. *Journal of Toxicology and Environmental Health, Part A*, *72*(2), 60–73.

42. Tan, A., Yildirimer, L., Rajadas, J., De La Peña, H., Pastorin, G., & Seifalian, A. (2011). Quantum dots and carbon nanotubes in oncology: A review on emerging theranostic applications in nanomedicine. *Nanomedicine*, *6*(6), 1101–1114.

43. Vashist, A., Kaushik, A., Vashist, A., Sagar, V., Ghosal, A., Gupta, Y. K., & Nair, M. (2018). Advances in carbon nanotubes–Hydrogel hybrids in

nanomedicine for therapeutics. *Advanced Healthcare Materials*, *7*(9), 1701213.

44. Wick, P., Manser, P., Limbach, L. K., Dettlaff-Weglikowska, U., Krumeich, F., Roth, S., & Bruinink, A. (2007). The degree and kind of agglomeration affect carbon nanotube cytotoxicity. *Toxicology Letters*, *168*(2), 121–131.
45. Wu, H. C., Chang, X., Liu, L., Zhao, F., & Zhao, Y. (2010). Chemistry of carbon nanotubes in biomedical applications. *Journal of Materials Chemistry*, *20*(6), 1036–1052.
46. Yang, D., Gao, S., Fang, Y., Lin, X., Jin, X., Wang, X., & Shi, K. (2018). The π–π stacking-guided supramolecular self-assembly of nanomedicine for effective delivery of antineoplastic therapies. *Nanomedicine*, *13*(24), 3159–3177.
47. Zare, H., Ahmadi, S., Ghasemi, A., Ghanbari, M., Rabiee, N., Bagherzadeh, M., & Mostafavi, E. (2021). Carbon nanotubes: Smart drug/gene delivery carriers. *International Journal of Nanomedicine*, *7*(4), 1681–1706.
48. Zhang, X., Yu, W., Guo, Y., Li, S., Chen, Y., Wang, H., & Bian, Z. (2023). Recent advances in photo electrocatalytic advanced oxidation processes: From mechanism understanding to catalyst design and actual applications. *Chemical Engineering Journal*, *455*, 140801.
49. Zhang, Y., Petibone, D., Xu, Y., Mahmood, M., Karmakar, A., Casciano, D., & Biris, A. S. (2014). Toxicity and efficacy of carbon nanotubes and graphene: The utility of carbon-based nanoparticles in nanomedicine. *Drug Metabolism Reviews*, *46*(2), 232–246.
50. Zhou, M., & He, J. (2007). Degradation of azo dye by three clean advanced oxidation processes: Wet oxidation, electrochemical oxidation and wet electrochemical oxidation—A comparative study. *Electrochimica Acta*, *53*(4), 1902–1910.
51. Zuo, G., Kang, S. G., Xiu, P., Zhao, Y., & Zhou, R. (2013). Interactions between proteins and carbon based nanoparticles: Exploring the origin of nanotoxicity at the molecular level. *Small*, *9*(9–10), 1546–1556.

CHAPTER 8

SOCIAL AND ETHICAL IMPLICATIONS OF NANOTECHNOLOGY

LEARNING OBJECTIVES

At the end of this chapter, readers will be able to:

1. Comprehend the applications of nanotechnology and their societal impacts
2. Understand the concept of technological convergence and how nanotechnology integrates with biotechnology, information technology, and cognitive science
3. Gain knowledge about the major socio-economic trends of nanotechnology
4. Recognize key ethical principles that should guide the development and deployment of nanotechnology

INTRODUCTORY EXAMPLE

IBM, a leading technology company, has been at the forefront of integrating nanotechnology into various fields, including medicine. One significant development is their work on using nanoparticles for targeted drug delivery systems. This innovation aims to improve the efficacy of treatments while minimizing side effects, particularly in cancer therapy.

IBM Research developed a nanomedicine approach that involves designing nanoparticles that can carry drugs directly to cancer cells. These nanoparticles are engineered to recognize and bind to specific markers on the surface of cancer cells, releasing the drug payload only when inside the target cells. This targeted delivery system ensures that healthy cells are largely unaffected, reducing the adverse effects typically associated with chemotherapy.

While the potential benefits of IBM's nanotechnology in medicine are substantial, including improved patient outcomes and reduced healthcare costs, there are also significant social and ethical considerations. Issues such as patient consent, data privacy, and long-term effects on human health need careful consideration. Furthermore, there is a risk that such advanced treatments could be expensive, exacerbating healthcare inequalities between different socio-economic groups.

This example illustrates the dual-edged nature of nanotechnology applications. On one hand, they offer groundbreaking solutions to pressing medical challenges; on the other, they raise important ethical and societal questions that must be addressed to ensure responsible development and equitable access.

UNIT INTRODUCTION

Science and engineering are just starting to investigate the possibilities for exploration and innovation at the nanoscale. However, there are already some extreme opinions advocating for government restrictions or complete prohibition of nanotechnology. Engaging in such action would be highly premature, as humanity is currently in the early stages of a lengthy research journey. It is necessary to fully understand the technical feasibility and practicality of nanotechnology, as well as its potential real-world applications (Suchman, 2001). Additionally, scientists must carefully consider and address any potential hazards that may arise from these applications, both directly and indirectly, before taking any further steps. Several writers in the mainstream media often portray nanotechnology as a modern version of Frankenstein's creature. However, it is important to highlight that even after almost two centuries since Mary Shelley's creation, biological science has not yet achieved the ability to bring such a creature to life (Bainbridge, 2004).

An effective approach to addressing social and ethical concerns begins with a precise description of the topic. However, there is a discrepancy between the definition commonly understood by the general public and the one used by specialists in this field. Merriam-Webster's Collegiate Dictionary provides the following definition for nanotechnology (Lopez, 2003). "The practice of controlling materials at an atomic or molecular level, particularly to construct minuscule devices such as robots." According to About.com, nanotechnology is the process of creating and utilizing gadgets that are only a few nanometers in size. However, it is important to mention that these definitions specifically emphasize nanoscale robot-like devices and exclude the main subjects discussed in most chapters of this volume (Schummer, 2006). The U.S. Government's National Nanotechnology Initiative offers a comprehensive definition of nanotechnology as research and technology development conducted at the atomic, molecular, or macromolecular levels within the 1–100 nm range. The purpose of this research is to gain an extensive knowledge of events and materials at the nanoscale and to design and utilize structures, devices, and systems with unique properties and functions due to their small or intermediate size (Sparrow, 2020). The novel and distinguishing characteristics and functions are formed within a crucial size range of matter, usually less than 100 nm. Nanotechnology research and development involves the deliberate manipulation and control of structures at the nanoscale, as well as their incorporation into bigger material components, systems, and designs (Doering, 2001).

Although the term "devices" is mentioned in this definition, it also encompasses materials, systems, and networks. Importantly, it predicts that numerous nanotechnology advancements will be incorporated into "bigger material components, systems, and architectures." Consequently, it is imperative to comprehend the social and ethical consequences of the larger-scale systems in which nanoscale structures are merely constituent parts (Tenner, 2001). Nanoscale engineering will likely have extensive applications in all areas of engineering and most areas of science. As a result, it will facilitate or enhance the impact of other technologies. Nanotechnology is a multifaceted field that cannot be categorized as solely ethical or unethical. Instead, it encompasses a multitude of diverse elements, including several nanotechnologies, each with a wide array of potential applications and potential misapplications when combined with other innovations and used for different objectives (Sykes & Matza, 2017).

Some individuals may infer from this line of thinking that the ethical and social repercussions of nanotechnology are not a cause for concern, as they will be overshadowed by the consequences of bigger systems. Some individuals may argue that it is premature to examine the societal consequences of nanoscale engineering, as it is impossible to analyze the effects of a technology that has not yet been developed (Berne & Schummer, 2005).

Nevertheless, the mere existence of writers advocating for the control or prohibition of nanotechnology indicates that the discussion over its worth to humanity has commenced, and it is imperative to contribute the most superior intellect and knowledge to this discourse. Furthermore, a handful of comparatively uncomplicated yet noteworthy applications have started to emerge, therefore allowing for the observation of an influence in those specific instances. Researchers from several disciplines, including sociologists, scientists, theorists, and engineers, have started to explore the future of nanotechnology (Viseu & Maguire, 2012).

Did You Know?

Some socks use silver nanoparticles to kill bacteria and prevent bad odors, making them stay fresh for longer periods.

8.1. APPLICATIONS AND SOCIETAL IMPACTS

Learning Objectives

- Comprehend the applications of nanotechnology in diverse fields
- Understand the societal impacts of nanotechnology in different applications

In 2000, the National Science and Technology Council organized a meeting at the National Science Foundation, resulting in the publication of a study titled "Societal Implications of Nanoscience and Nanotechnology." A group of 64 experts from university research, governmental research facilities, and corporate research have compiled a comprehensive list of potential nanotechnology uses (Bernstam, 1986). Some of these applications are already being produced, while many more are predicted to become feasible only after several decades of research. The workshop participants provided a range of examples of potential uses across different parts of the economy and areas of life (Abdullah et al., 2022).

i. Enhanced equipment for the semiconductor industry, such as integrated circuits that incorporate transistors made from carbon nanotubes. Nanostructured catalysts are utilized in chemical production and in more efficient converters to address pollutants from automotive exhausts and other forms of combustion.

ii. Develop mass amounts of nanomaterials that are both lighter and stronger, to enhance the safety and efficiency of vehicles (Pulido et al., 2020).

iii. Enhanced pharmaceuticals with capabilities such as targeted distribution to specific locations like tumors and the utilization of non-water soluble compounds as treatments.

iv. Affordable and dependable filters for the purification of water, removal of salt from water, control of pollutants from industries, and purification of air.

v. Enhancing the efficiency of solar energy conversion, therefore decreasing the existing dependence on oil and providing an attractive alternative to nuclear energy to meet future power requirements (Gupta et al., 2018).

vi. Development of highly efficient fuel cells and advanced hydrogen storage systems, resulting in the production of environmentally friendly autos, trucks, and busses.

vii. Nanoparticle-reinforced polymers are being developed to create composite materials that are more robust and have longer lifetimes. These materials are specifically intended to function well in certain applications and minimize waste.

viii. Molecularly tailored degradable fertilizers and pesticides are developed specifically for precision farming. They are intended to be efficiently delivered to the targeted areas, while also preventing any undesirable side effects (Bornmann, 2013).

ix. Revolutionary space launch vehicles possessing high mass ratios, environmentally friendly yet potent engines, and little electric power re□uirements to ultimately fulfill the potential of space travel.

x. Nanoscale components are utilized in sensor systems to rapidly detect and distinguish pollutants, pathogenic organisms, and chemical or biological attack weapons. This enables prompt and suitable medical care or security measures.

xi. Coatings with the ability to confer exceptional features to common materials, such as self-cleaning windows and heat-reflecting external building surfaces (Cohen & Syme, 1985).

Well-informed scientists and engineers acknowledged that these and numerous more comparable applications were feasible advantages of nanoscale study, but they acknowledged that advancement comes with expenses. For instance, the emergence of a novel industry frequently leads to the decline or total elimination of a preexisting industry, as exemplified by the typical scenario of buggy-whip producers becoming obsolete due to the advent of automobiles (De Jong et al., 2014). It is necessary to evaluate the potential advantage of a new technology comprehensively, considering all the costs associated with its whole life cycle. For instance, an electric vehicle might mitigate certain types of pollution but may also contribute to others, such as the emission of lead or other hazardous compounds into the atmosphere during the production and disposal of the electric battery (Molas-Gallart & Tang, 2011).

Nanotechnology businesses will inevitably give birth to numerous ethical challenges related to fairness. For instance, it will be crucial to allocate resources towards implementing suitable measures to protect personnel involved in perilous production procedures. Typically, scientific findings are not eligible for patenting, whereas novel goods and manufacturing techniques can be patented. Conse□uently, the field of nanoscience, which is unconventional and without a well-defined boundary between discovery and invention, presents significant challenges in terms of safeguarding intellectual property. Frequently, it will be essential to amalgamate information and expertise from multiple sources to develop a financially lucrative application, leveraging government-funded higher education research initiatives and various specialized industrial corporations, necessitating enhanced prototypes for cooperation across enterprises (Roco & Bainbridge, 2007).

According to the Societal Implications report, the National Nanotechnology Initiative (NNI) has a substantial impact on the general public, which needs to be addressed through public participation and the inclusion of social scientists. Incorporating a diverse array of interests, attitudes, and viewpoints is crucial when making decisions that shape the future advancement of nanotechnology. Engaging people in society or their elected officials provides the additional advantage of acknowledging their interests and securing their backing (Baumgartner, 2008). Integrating sociologists and humanist researchers, such as philosophers specializing in ethics, into the process of establishing goals for nanotechnology is a crucial measure for the NNI. As scientists or diligent academics themselves, they can appreciate the professional honesty of nanoscientists and nanotechnology researchers, while offering a novel viewpoint. If provided with adequate assistance, they could acquire a profound understanding of specific nanotechnology, enabling them to conduct a thorough and informed assessment. Simultaneously, they possess professional training as agents of the public interest and can serve as intermediaries between nanotechnology researchers and the public or the government. Their contribution can assist in optimizing the societal advantages of technological advances while minimizing the likelihood of significant public disputes (Floridi, 2010).

Utilizing sociological and economic studies can assist producers and the government in making informed judgments regarding the implementation of new technologies, hence optimizing their advantages for humanity. Moreover, conducting research that demonstrates technical proficiency in analyzing the societal consequences of nanotechnology will assist legislators and members of the public in forming an accurate understanding devoid of unfounded optimism or anxieties. According to the paper, the National Nanotechnology Initiative is starting now, which means there is a unique chance to incorporate societal research and discussions right from the start and make them an essential component of the NNI investment plan (Bainbridge, 2010).

The workshop formulated several proposals for the government, education, and the private sector on how to effectively handle the societal and ethical aspects of nanotechnology. Below are several fundamental ideas in the fields of higher education, social sciences, and ethics:

i. It is essential to incorporate nanoscale concepts into scientific and engineering courses at all levels. This will ensure that a broad variety of students gain a comprehensive grasp of the topic, while also establishing intellectual connections between nanotechnology and several other disciplines.

ii. Nanotechnologists should get training that encompasses societal ramifications and ethical sensitivity, to ensure that their future work is guided by principles that aim to optimize human benefit.

iii. An adequate number and diverse range of economic and social scientists should undergo comprehensive interdisciplinary training to ensure their readiness for employment in the field of nanotechnology.

iv. Formal measurement tools, such as economic and social indicators, should be created and continuously used to accurately track the significant changes in business, schooling, and social services as they happen.

v. Government departments, nonprofit organizations, and industry should provide financial backing for a diverse array of top-notch, scientific studies on nanotechnology. These studies should be grounded in solid theories and should investigate both the decision-making processes that influence the development of this emergent technology, as well as its specific effects on society once it is put into use.

vi. To assess the likely prospective intellectual and societal effects of nanotechnology and nanoscience in all three timelines, it is necessary to establish an understanding and institutional framework (Bordogna & Albano, 2007).

vii. The government and commercial sector should provide efficient means of immediately alerting the public about emerging concepts, projects, potential uses, and ethical concerns.

viii. It is necessary to establish formal organizations and methods to guarantee the involvement of various society organizations and the public at large in determining research and development priorities, as well as delivering timely input for decision-makers.

ix. Government and business should collaborate to create managerial strategies and guidelines that can efficiently integrate all pertinent information and valid stakeholder concerns (Benoit-Norris et al., 2012).

Implementing these proposals will re□uire a significant investment of time and effort, although significant advancement has already been achieved. Rosalyn Berne at the University of Virginia has received funding from the National Science Foundation to conduct a narrative evaluation of discourse from scientists and engineers in the field of nanotechnology, to study the evolving ethics of this field. Additionally, Michael Gorman and his team at the same institution have been funded to investigate the possibilities of multidisciplinary research on the social and ethical aspects of nanotechnology (Wilhelm et al., 2015).

8.2. TECHNOLOGICAL CONVERGENCE

Learning Objectives

- Understand the concept of technological convergence
- Comprehend how nanotechnology integrates with biotechnology, information technology, and cognitive science

Following the Societal Implications workshop, the National Science Foundation (NSF) and the U.S. Department of Commerce organized a larger meeting to explore the possibilities of nanotechnology merging with three other influential scientific and engineering disciplines: information technology, biotechnology, and cognitive science. These four fields are collectively referred to as "NBIC." The report, titled "Converging Technologies for Improving Human Performance," analyzed the advantages of technology convergence in enhancing human capabilities to meet their requirements and accomplish their lawful objectives (Laurent & Petit, 2005).

The convergence of various NBIC technologies will rely on the inherent unity of materials that exist at the nanoscale and the integration of technology at that level. This scale is crucial for the functioning of future information technology and biotechnology hardware. It is also the scale at which basic investigations on the nervous system of humans will be conducted (Fisher, 2007). The integration of formerly distinct scientific disciplines will give rise to essential tools for modifying key elements, such as studies and production procedures. By utilizing advancements in mathematics and computers, it will become feasible to create models of complex, interconnected, and hierarchical systems that encompass phenomena ranging from the nano to the cosmic scale (Ronteltap et al., 2011).

The Converging Technologies researchers, engineers, and politicians identified six broad domains that are expected to yield human advantages and raise social concerns:

a. The overarching potential of merging technologies
b. Enhancing human cognitive abilities and facilitating communication
c. Enhancing human well-being and physical capacities
d. Improving group and societal results
e. National security
f. Integrating science and education (Roco, 2005).

A multitude of valuable applications were outlined, with varying levels of detail, that may plausibly be anticipated to arise from a decade or two of intensive research and development. These concepts rely on collaborations between nanotechnology and any or all of the other NBIC fields:

i. The utilization of comfortable and portable sensors and processors will augment individuals' consciousness regarding their health status, the surrounding environment, chemical contaminants, potential dangers, and pertinent information about local enterprises, natural resources, and similar matters of interest (Laurent, 2007).
ii. The human body will exhibit enhanced endurance, improved health, increased energy levels, greater ease of healing, and heightened resistance to various forms of stress, biological hazards, and aging mechanisms.
iii. Machines and buildings, ranging from residences to airplanes, will be built using materials that possess precisely the needed characteristics, such as the capacity to adjust to varying circumstances, exceptional energy performance, and environmental compatibility (Adams et al., 2018).

iv. By utilizing a blend of advanced technologies and therapeutic interventions, numerous physical and mental infirmities can be effectively mitigated, and certain handicaps that have afflicted millions of individuals can be eliminated.

v. The enhancement of national security will be significantly bolstered through the utilization of lightweight, data-intensive warfighting systems, autonomous combat vehicles, versatile intelligent materials, impervious networks of data, advanced intelligence-gathering systems, and efficient countermeasures against chemical, biological, radiological, and nuclear attacks.

vi. The capacity to manipulate the genetic makeup of people, animals, and plant life will significantly enhance human well-being, while concurrently establishing a broad agreement on legal, ethical, and moral concerns (Prigulnyy, 2012).

vii. The agriculture and food business will significantly enhance productivity and minimize waste by implementing cost-effective, intelligent sensors that continuously monitor the status and requirements of crops, livestock, and agricultural goods.

viii. The transportation system will be secure, affordable, and rapid thanks to the widespread availability of real-time information systems, highly efficient vehicle designs, and the utilization of synthetic materials and nanoscale machines for optimal performance.

ix. The work of scientists will undergo a revolutionary transformation by incorporating methodologies and techniques developed in other scientific disciplines. For instance, genetic research will benefit from the application of principles derived from natural language processing, while cultural research will be enhanced by the utilization of principles derived from genetics.

x. A unified and diverse curriculum, based on a comprehensive and hierarchical intellectual paradigm, will revolutionize formal education. This curriculum will focus on understanding the structure of the physical world, ranging from the smallest scale of nanotechnology to the largest size of the universe (Bainbridge & Roco, 2016).

Many of these application domains rely on the advancement of nanoscale elements for sensor, computing, and communication networks. Other approaches entail a harmonious collaboration between the manipulation of inorganic materials at the nanoscale and the fields of chemical science or biology. Some experts predict that in the future, it will be feasible to construct enormous structures that are designed at the nanoscale to improve the characteristics of construction, automobiles, and even the human body. At the core of all applications is a fundamental revolution that integrates research and offers people a full understanding of nature and technologies (Chou & Liou, 2011).

The workshop did not include nuclear science and atomic engineering in its examination of the cluster of NBIC converging technologies. Therefore, the following discussion provides additional information to complement its results. Scientists had already comprehended in the early 1940s that there are essentially two methods by which a rogue nation can amass its stockpile of weapons-grade nuclear fuel. One is using uranium as fuel in a nuclear reactor to produce plutonium, causing a transformation of the naturally occurring uranium isotope with an atomic weight of 238 (Wetter, 2006). However, it is challenging to hide this information from other countries, particularly when nuclear reactors are subject to international scrutiny. Another method involves the isolation of the comparatively scarce uranium isotope U_{235}, which, similar to plutonium, is appropriate for utilization in weaponry. The initial technique employed for separation was gas diffusion; however, it was later replaced by the centrifuge method. Additionally, alternative methods such as lasers or electromagnetic separation have been explored (Koepsell, 2011).

The gas diffusion method involves the passage of uranium hexafluoride, a chemical compound containing uranium, through a porous barrier, such as a ceramic wall with minuscule apertures. U_{235} molecules traverse the barrier at a somewhat higher velocity compared to U_{238} molecules. This method is highly inefficient due to the similarity between the isotopes, necessitating numerous passes through a barrier to properly enrich the proportion of U_{235} for nuclear weapon usage (Song et al., 2017). An alternative method involves the utilization of centrifuges, which are high-speed rotating devices that employ uranium hexafluoride to differentiate the two isotopes based on their minute disparity in mass. Both approaches are highly challenging and expensive, nevertheless. Possibly, nanotechnology has the potential to somewhat decrease the level of difficulty by creating porous barriers with precisely calibrated holes and by offering centrifuge rotors that are both lighter and stronger, capable of operating at increased speeds. Thus, in the absence of a thorough examination, one may hastily infer that nanotechnology poses a threat by enabling the widespread production of nuclear weapons (Kim, 2011).

Just a small number of nano goods may require regulation as "controlled substances," similar to narcotics, pesticides, asbestos, and antibiotics. Possibly, porous barriers with the precise pore size for separating uranium isotopes might be regulated in some manner. Certain items can be manufactured and utilized under stringent controls, while a select number may be completely prohibited. The majority would not necessitate any form of limitations (Bae et al., 2013).

The expenses associated with excessive legislation would be exceedingly elevated, causing harm to the same individuals it was originally designed to safeguard. Nanoscale porous barriers are expected to be highly beneficial in the early stages of nanotechnology. These applications encompass a broad spectrum, ranging from the purification and desalination of water to the segregation of important compounds in biotechnology manufacturing procedures. Likewise, advanced materials that possess exceptional strength yet are lightweight may find numerous uses in various civilian industries. Some of the things that would be advantageous to a large number of individuals are lightweight yet durable eyeglass frames, protective helmets, and manual tools (Gelfert, 2012).

Remember

Responsible research, public engagement, and adherence to ethical principles are crucial for the socially responsible development of nanotechnology.

8.3. MAJOR SOCIO-ECONOMIC TRENDS

Learning Objectives

- Comprehend how nanotechnology relates to key socio-economic wider trends going on in the world.

The importance of nanotechnology is determined not only by its own achievements and its integration with additional technologies but also by how these strictly technological advancements align with broader global trends (Roiko et al., 2012). Currently, numerous tendencies in the social sciences and government policy spheres are muddled and provoke significant discussion. Attempting to conclude future patterns from a certain moment in history is always risky. However, the current period seems extremely disordered, to the extent that the occurrences of only one day could significantly alter the course of humanity (Yadav et al., 2013).

Several years ago, there was a widespread agreement that the significant increase in the global population, sometimes referred to as the population boom, was the primary quantitative trend that had to be considered in any forecasts. However, by the mid-1980s, it became evident that population explosion had been substituted by the peril of population reduction in the majority of advanced industrial nations, except possibly the United States (Masara et al., 2021).

Prior to the onset of declining birth rates, the impacted nations experienced a rise in median age and pensions expenses, a drop in social dynamism and inventiveness, and governments faced challenges in effectively addressing the financially demanding services. This is one of the obvious explanations why the US has started to diminish the impact of its European allies and Tokyo (Sahoo, 2010).

While nanotechnology may not have a clear and direct correlation with fertility rates, two possible indirect linkages can potentially offset each other. The rapid advancement of nanotechnology has the potential to rejuvenate economic growth and foster technical innovation, somewhat mitigating the adverse effects of declining birth rates. Nevertheless, the reluctance to embrace change that characterizes countries with a large number of older individuals may hinder the intellectual curiosity and willingness to take risks necessary for the advancement of nanotechnologies and convergence in technology (Bürgi & Pradeep, 2006).

An associated phenomenon is the continual enhancement of well-being stemming from advancements in medical research and the economic expansion re□uired to finance progressively expensive healthcare. However, there is no guarantee of continued progress in health. Throughout the 19th century, there is evidence to suggest that the overall health of Americans may have experienced a notable deterioration for several decades, even during economic advancement. The indisputable enhancement in health throughout the twentieth century is sometimes ascribed to medical advancements, although health awareness and public cleanliness may have had a more significant role (Aidukaite, 2011). The U.S. Centers for Disease Control contends that significant enhancements in health and lifespan can be attained by lifestyle modifications, specifically by increasing physical activity, limiting dietary fat intake, and abstaining from smoking. The advent of antibiotics has significantly contributed to the extension of the human lifespan, while advancements in cardiology have successfully rescued several individuals from early death caused by heart attacks. However, the progress in combating AIDS, cancer, and most age-related diseases has been remarkably sluggish. Overall, economists conclude that the rising investments in healthcare are yielding positive results, although not uniformly across all sectors or with significant advantages (Arumugam et al., 2023).

There are multiple reasons to have a negative outlook on the future advantages of healthcare unless there are significant scientific advancements. Currently, no cure has been discovered for any persistent viral infections, with AIDS being the most widely known instance. Bacteria are exhibiting a fast adaptation to antibiotics, while the pace of new prescription drug development is diminishing. The mean life span of a Caucasian American male born in 1900 was 48 years, however by 2000, this life expectancy had risen to 74 years (Talala et al., 2009). The mean lifespan for white females has risen from 51 to 80. Extrapolating these statistics with the same growth rate implies that the average lifespan for males in 2100 may be 114 years, while for females it could be 125 years. Nevertheless, based on more accurate assumptions, the U.S. Census Bureau has forecasted that the life expectancy of those born in the year 2100 in the United States might potentially be approximately 88 years for males and 92 years for females. This signifies advancement, but at a rapidly decreasing pace, and it relies on the continuation of economic expansion (Bhattacharyya et al., 2011).

Moore's Law, observed by Gordon Moore, states that the density of transistors on the most modern microprocessor doubles approximately every 18 months, making it the most profitable technology trend in recent years. The swift advancement of computer and communications hardware has played a crucial role in enabling the adoption of novel software programs, networking, and information technology as a whole (Gajdzik et al., 2021). These factors, in turn, have played a significant role in driving economic expansion. Nevertheless, the semiconductor industry is reaching the physical boundaries of conventional techniques used to manufacture integrated circuits, particularly computer chips. Consequently, the progress predicted by Moore's Law will come to a halt unless there are significant advancements in nanoscale technologies, such as molecular logic gates and nanotube transistors (Sarma & Chaudhury, 2009).

It might be argued that many significant advancements in technology during the 20th century have already come to a halt. Except for the enhanced capabilities of scientific e□uipment to gather and analyze astronomical data, the progress of the space program has been little since the mid-1970s. Civilian aviation and transport on land have made only incremental advancements in design aspects since the 1950s (Ponkratova et al., 2021). Despite extensive endeavors, minimal advancements have been achieved in the regulation of nuclear fusion for power generation. The progress of nuclear fission energy production has been significantly impeded by public debates, and there have been no significant advancements in ensuring the safe operation of reactors or the elimination of radioactive wastes that have alleviated the public's fears (Tegart, 2004).

If technological advancement were to slow down or come to a standstill, the entire economy would be significantly impacted. Initially, the advanced industrialized countries would relinquish the superiority they presently possess over emerging ones. For instance, the American semiconductor sector would likely face a significant decline if it were to lose its technological dominance, as countries with lower labor costs could manufacture similar products at a cheaper expense (Akkam & Khatatbeh, 2021). Technological products, ranging from computer chips to medications, would be transformed into commodities, produced in locations where the cost is the lowest. Advanced nations' governments would confront budgetary crises of unparalleled size due to their substantial pension obligations and expensive social services. Undoubtedly, the economies of these countries have the potential to plummet swiftly to the global average, giving rise to societal unrest that may result in unforeseeable acts of violence (Gupta et al., 2013).

Practice Problem

Discuss the importance of the precautionary principle in nanotechnology. How can it be applied to mitigate potential risks?

Solutions to Practice Problem

The precautionary principle emphasizes caution in the development and deployment of new technologies to prevent harm when scientific understanding is incomplete. In nanotechnology, this principle is vital due to the unknown long-term effects and potential risks associated with nanoscale materials.

Application of the Precautionary Principle:

- **Risk Assessment**: Conduct thorough risk assessments before approving new nanotechnology applications
- **Regulations**: Implement strict regulatory frameworks to oversee the use of nanomaterials in consumer products
- **Research**: Encourage extensive research on the environmental and health impacts of nanomaterials
- **Public Involvement**: Involve stakeholders, including the public, in decision-making processes to ensure diverse perspectives are considered

8.4. SOURCES OF ETHICAL BEHAVIOR

Learning Objectives

- Understand the origins of ethical behavior in the field of nanotechnology
- Comprehend the strain theory, control and subculture theory for ethical behavior in the field of nanotechnology

The ethical concerns of nanotechnology extend beyond deliberate acts of harm against others, encompassing the responsibility to prevent any unforeseen negative outcomes that may arise. An illustration of the former may be a weapon, whereas an instance of the latter could be a sort of environmental pollution (Lu et al., 2012). Regardless of the situation, the potential harm could be ethically justifiable. This could occur when a nation utilizes a weapon to protect itself from an assault, or when a small amount of pollution in the environment is balanced by significant advantages for humanity. Additional ethical concerns arise with the potential harm inflicted on nanotechnology owners. This includes instances where a rival corporation infringes upon a nanotech patent or when a government prohibits a nano-related item without conducting a thorough evaluation of its scientific merits (Corley et al., 2016). Journalists or famous writers who disseminate incorrect data about nanotechnologies may also be engaging in unethical behavior, whether they are intentionally deceiving or simply neglecting to exercise professional diligence in fact-checking (McComas, 2012).

It is crucial to acknowledge that there are both sins of omission and sins of action. Not being able to create a useful nanotechnology application could likewise be considered unethical. Consider a scenario where a laboratory has made a breakthrough in utilizing nanotechnology to enhance the efficacy of chemotherapy in treating cancer. However, the company that holds the patents deliberately obstructs its development due to the higher profitability of their less operational traditional chemotherapy products. This action could be seen as e□ually unethical to inflicting intentional harm (Grunwald, 2020).

For more than a century, researchers in social sciences have been investigating the sources of ethical behavior, as well as its counterpart, unethical behavior, in disciplines such as group social psychology, deviant sociology, and criminology. An individual's behavior is influenced by a multitude of circumstances, many of which cannot be precisely measured, and randomness appears to have a notable impact on human events. Hence, the conclusions drawn from social science research typically exhibit a statistical or conditional character. Despite their limitations, probabilities can nevertheless be utilized to formulate sensible laws, government programs, and investment strategies instead of relying solely on certainties (Eosco et al., 2014). The academic work on the topic is extensive, comprising numerous articles going back to the 1840s. It encompasses various schools of thought and uses unique terminology. To achieve the current objectives, it is necessary to condense the extensive and disorganized body of intellectual knowledge to its most significant and crucial elements. One effective beginning consists of the following collection of four concepts, each of which can be articulated as a theoretical explanation for why individuals may engage in ethical rule violations (Jung, 2009):

i. **Learning**: An individual breaks a rule because they have acquired knowledge that engaging in such behavior is advantageous.

ii. **Strain**: An individual breaches a regulation due to their frustration in trying to meet societal expectations.

iii. **Control**: An individual breaches a regulation due to a deficiency of established social connections that could otherwise deter such behavior.

iv. **Subculture**: An individual breaches a rule due to their affiliation with an organization or network that actively promotes rule-breaking (Yamin, 2020).

Initially, when these ideas were introduced in the social scientific literature, the supporters of the theory often portrayed it as the whole explanation for deviant conduct. However, each of these four ideas has value, and an accurate explanation typically involves a combination of two or more. Learning Theory is occasionally understood via an economic or "rational choice" lens, suggesting that individuals will engage in actions they consider to be advantageous. However, the entirety of human existence is not dedicated just to engaging in commercial transactions, where individuals can assess pricing and seek advantageous bargains. Instead, decision-making occurs in various distinct situations. Irrespective of the monetary value of options, individuals tend to depend on their personal past experiences when choosing between recognizable courses of action (Kalshoven et al., 2011).

If the executives of an industrial firm have discovered that they may attain greater financial gains by disregarding legislation about environmental pollution, and if they have not faced substantial penalties from the government as a consequence, they will be more inclined to engage in future acts of pollution. Conse□uently, a firm that is found responsible for contaminating the environment using traditional methods is likely to persist in its pattern of pollution when it implements new nanotechnologies. This implies that implementing and enforcing stringent rules against contamination, in general, can be an effective approach to minimize the probability of industrial businesses causing pollution when they embrace new nanotechnologies. Such regulations would compel these corporations to learn and adopt practices that prevent pollution (Granitz, 2003).

According to Strain Theory, an individual deviates from societal norms due to their frustration in trying to attain one of the values that society promotes. This concept has been employed to elucidate the reasons behind the inclination of individuals from impoverished communities or marginalized groups to resort to criminal activities as a means to acquire wealth and social standing. However, it can also shed light on the motivations behind a large corporation resorting to fraudulent accounting procedures when its actual business growth stagnates. The main concept is that when a person or organization becomes dedicated to a specific type of success that is endorsed by society – such as wealth, political influence, widespread recognition, or prestige within a specific social sphere – they may encounter frustration in achieving that success, which in turn drives them to engage in unethical behavior (Mitchell et al., 2020).

Possibly, a nanotechnology firm may experience strain if it has dedicated itself to a novel and untried range of goods that ultimately prove to be less efficient, more detrimental, or more expensive than initially anticipated. The corporation has invested not just its financial resources and labor, but also its public reputation in pursuit of success, but achieving success may prove elusive. Subsequently, the company may engage in unethical practices by falsely promoting its products as more efficacious than they truly are, withholding information about their potential harm to people's health or the ecosystem, and manipulating its financial statements to mislead investors. Indeed, some conventional companies have already undertaken this action, and, paradoxically, they occasionally resolve their issues before the public's awareness, thus regaining their ethical standing and achieving success once more (Downe et al., 2016).

In 1938, Merton introduced the most influential form of Strain Theory, in which he categorized three types of adaptation to three escalating levels of strain. If an individual is unable to meet society's standards according to traditional norms, they may feel frustrated and resort to breaking those rules to pursue the desired value. This behavior, as termed by Merton, is referred to as innovation. If this approach proves ineffective, an individual may opt to replace both the prevailing values and norms of society with alternative ones, a phenomenon referred to as rebellion. However, if this attempt proves unsuccessful, the individual may forsake their principles, as well as societal standards, and descend into a disheartened condition of passivity known as retreatism. Retreatism has little significance, but successful rebellion often leads to the radicalization of society. However, innovation, as defined by Merton, can foster scientific and technological advancements. Furthermore, Merton played a significant role in establishing the field of social science and technology. However, he did not fully explore the consequences of the Strain Theory in relation to science. Moreover, his unique notion of innovation was primarily employed to elucidate the criminal actions of impoverished individuals who transgressed the law in pursuit of the values, such as monetary gain, upheld by capitalist society (Van Wart, 1996).

Throughout the history of science, there are numerous instances where the Strain Theory can be applied to elucidate scientific and technological breakthroughs that were advantageous rather than unethical. However, there are also cases where the outcome was a combination of benefits and drawbacks, suggesting that better management could have prioritized the benefits. During the 17th century, England emerged as a global leader in scientific advancement, possibly as a result of a nationalistic response to its comparative disadvantage in terms of population and resources compared to Spain and France. During the latter part of the 19th and early 20th centuries, Germany did not have the same level of access to natural resources as the powerful colonial nations, such as Britain and France. The perceived lower quality of German chemistry may have served as a catalyst for significant advancements, including Adolf Bayer's creation of indigo dye and the nitrogen fixation techni□ue pioneered by Fritz Haber and Karl Bosch. This breakthrough enabled Germany to sustain its production of explosives during World War I, even after losing access to imported nitrates. The Treaty of Versailles, which marked the end of World War I, imposed restrictions on German long-range artillery. In response, the Germans created the V-2 rocket as a means to circumvent this limitation, thus paving the way for the era of space exploration (Ma'Ayan & Carmeli, 2016).

In the field of Control Theory, individuals deviate from societal standards due to a deficiency in enduring social connections that may otherwise deter such behavior. This is a highly effective scientific elucidation of numerous prevalent criminal activities, such as petty theft, and many types of offenders tend to exhibit a deficiency in strong friendships or familial connections that could potentially curtail their aberrant conduct. The relevance of Control Theory to nanotechnology ethics may not be immediately apparent due to the absence of any single human with the authority to make significant decisions regarding new uses. The hypothesis aligns seamlessly with the stereotypical portrayal of Dr. Frankenstein in Hollywood, who resides as a recluse in an isolated castle, granting him the freedom to carry out any gruesome experiment he desires. However, it is important to note that no genuine scientist adopts such a lifestyle (Magalhães et al., 2019). Nevertheless, a modified version of Control Theory posits that an organization or community may experience a condition known as social disorganization, characterized by a widespread but incomplete form of isolation. This condition arises when communication channels are inadequate and a significant number of individuals exhibit a moderate level of deviation due to their partial detachment from one another. In a research laboratory characterized by social disorganization, scientific integrity may be compromised, failing to disseminate crucial information to the scientists who require it. Additionally, the management may develop inaccurate perceptions of the qualities of the nanostructured material. This has the potential to trigger a cascade of unethical conduct, for which no single individual can be predominantly held responsible. For example, the outcome might involve a corporate resolution to manufacture a novel type of airplane component that was prone to unforeseeable catastrophic malfunction, resulting in accidents and loss of human lives. In this scenario, the unethical conduct is fragmented into numerous little instances, perpetrated by multiple persons, whenever one of them presents a □uestionable finding as a scientific fact. To restore the truth-focused professional standards of science, it would be most effective to establish effective channels of interaction and cooperation, reconnecting researchers with one another and with the scientific community (Rosenbaum et al., 2004).

Subculture Theory and Control Theory share similarities, but differ in their focus: Subculture Theory examines group dynamics, while Control Theory focuses on individual behavior. An individual breaches a rule due to their affiliation with a subculture that not only opposes the rule but also promotes its violation. A subculture refers to a distinct group or network that possesses a unique set of rules, roles, and values that deviate from those of the larger culture in which it exists. At its most extreme, a subculture may exhibit a radical ideology or dogma. However, in reality, nearly every social group can be considered a subculture to some degree, even if just to a small level. Research conducted by social psychologist Janis has demonstrated that even highly influential groups have the potential to develop into deviant subcultures when they choose to separate themselves from the larger community, as exemplified by the Nixon White House during the Watergate crisis (Stevens, 1999).

Even if a research laboratory is operating effectively, the overall organization it belongs to can make detrimental judgments if the top management follows a subculture that practices "groupthink," which is the rejection of knowledge that contradicts the group's views and convictions, as described by Janis. Occasionally, this could lead to harm being inflicted upon others. For example, if a cutting-edge cancer treatment that utilizes nanotechnology were to be manufactured and distributed to patients in hospitals without first establishing safe protocols for its administration (Kump & Scholz, 2022). Alternatively, the organization itself may suffer negative consequences, such as if a semiconductor corporation were to make a decision to invest in molecular logic gates without thoroughly evaluating the possibilities of carbon nanotube transistors. Under such circumstances, one may argue that the corporation is deserving of the consequences it faces. However, there is an ethical concern with the company's obligation to its owners and employees, especially the scientists in the research laboratory who were prepared to do the necessary studies for an accurate judgment (Kantor & Weisberg, 2002).

As demonstrated by the example of Tension Theory, tension can occasionally lead to positive outcomes instead of negative ones. Learning Theory elucidates the reasons behind the formation of ethical innovation habits in individuals or organizations. Additionally, Control Theory and Subculture Theory demonstrate that individuals or groups may achieve optimal innovation when they are slightly disconnected from conventional thinking. Therefore, the problem lies not in avoiding these four components, but in determining the appropriate quantities of them to promote constructive innovation while preventing the occurrence of severe unethical conduct. The core dilemma in any community that values freedom is how to unleash human ingenuity while curbing greed, immorality, and inhumanity (Al Halbusi et al., 2021).

SUMMARY

- This chapter examines the wide-ranging uses of nanotechnology and its possible effects on society, emphasizing the merging of different technologies, significant socio-economic patterns, and the origins of ethical conduct in this domain.
- The chapter explores the various uses of nanotechnology in different fields, such as medical, electronics, energy, and environmental science. These applications possess the capacity to greatly enhance the quality of life.
- Technological convergence occurs when nanotechnology intersects with other growing fields such as biotechnology, information technology, and cognitive science. This convergence has the potential to result in groundbreaking discoveries, but it also presents intricate ethical and regulatory dilemmas.
- The chapter then delves deeper into significant socio-economic developments that are impacted by nanotechnology, including changes in labor markets, economic disparity, and global competitiveness. Nanotechnology's revolutionary potential can result in the displacement of jobs in existing industries, while simultaneously generating new opportunities in high-tech sectors.
- Ultimately, this chapter investigates the origins of moral conduct in the advancement and use of nanotechnology. Ethical issues encompass the use of the precautionary principle, the practice of responsible research and innovation, and the inclusion of a wide range of stakeholders in decision-making processes.

MULTIPLE CHOICE QUESTIONS

1. Which of the following areas is NOT a common application of nanotechnology?

a. Healthcare

b. Electronics

c. Agriculture

d. Astronomical Observation

2. Which ethical principle emphasizes caution in the deployment of new technologies to avoid harm?

a. Beneficence

b. Autonomy

c. The Precautionary Principle
d. Justice

3. How might nanotechnology contribute to economic inequality?

a. By providing high-paying jobs to everyone
b. By making advanced technologies more expensive
c. By creating a digital divide between those who have access to nanotechnology and those who do not
d. By reducing the cost of consumer goods

4. Which of the following is a potential risk associated with the environmental impact of nanotechnology?

a. Increased agricultural yields
b. Reduced energy consumption
c. Toxicity to aquatic life due to nanoparticle pollution
d. Enhanced renewable energy technologies

5. What role do policymakers play in the ethical development of nanotechnology?

a. They design nanomaterials for specific applications
b. They fund nanotechnology research exclusively
c. They create regulations and guidelines to ensure safe and ethical use of nanotechnology
d. They conduct all nanotechnology research

REVIEW QUESTIONS

1. What are some potential negative societal impacts of nanotechnology?
2. How does the convergence of nanotechnology with biotechnology and information technology create new opportunities and challenges?
3. How might nanotechnology influence economic ine□uality within and between countries?
4. What are the key ethical principles that should guide the development and use of nanotechnology?
5. What role do stakeholders, including policymakers, scientists, and the public, play in promoting ethical behavior in the field of nanotechnology?

Answers to Multiple Questions

1. (d); 2. (c); 3. (c); 4. (c); 5. (c)

REFERENCES

1. Abdullah, M., Madain, A., & Jararweh, Y. (2022). ChatGPT: Fundamentals, applications and social impacts. In *2022 Ninth International Conference on Social Networks Analysis, Management and Security (SNAMS)* (Vol. 2, pp. 1–8). IEEE.
2. Adams, T. L., Taricani, E., & Pitasi, A. (2018). The technological convergence innovation. *International Review of Sociology*, *28*(3), 403–418.
3. Aidukaite, J. (2011). Welfare reforms and socio-economic trends in the 10 new EU member states of Central and Eastern Europe. *Communist and Post-Communist Studies*, *44*(3), 211–219.
4. Akkam, Y., & Khatatbeh, M. (2021). Future ethical implications and potential health consequences of nanotechnology research in Jordan: A scient metric analysis. *Tropical Journal of Pharmaceutical Research*, *20*(2), 425–431.
5. Al Halbusi, H., Williams, K. A., Ramayah, T., Aldieri, L., & Vinci, C. P. (2021). Linking ethical leadership and ethical climate to employees' ethical behavior: The moderating role of person–Organization fit. *Personnel Review*, *50*(1), 159–185.
6. Arumugam, V. R., Ng, B. K., & Thambiratnam, K. (2023). A Delphi study on technical and socio-economic perspectives of nanotechnology and ICT industries relations. *Systems*, *11*(4), 190.
7. Bae, S. H., Lim, J. S., Shin, K. M., Kim, C. W., Kang, S. K., & Shin, M. (2013). The innovation policy of nanotechnology development and convergence for the new Korean government. *Journal of Nanoparticle Research*, *15*, 1–15.
8. Bainbridge, W. S. (2004). Social and ethical implications of nanotechnology. *Springer Handbook of Nanotechnology*, *4*(2), 1135–1136.
9. Bainbridge, W. S. (2010). Governing nanotechnology: Social, ethical, and human issues. *Springer Handbook of Nanotechnology*, *7*(3), 1867–1883.
10. Bainbridge, W. S., & Roco, M. C. (2016). Science and technology convergence: With emphasis for nanotechnology-inspired convergence. *Journal of Nanoparticle Research*, *18*, 1–19.
11. Baumgartner, C. (2008). Bionanotechnology: A new challenge for

ethical reflection? *Emerging Conceptual, Ethical and Policy Issues in Bionanotechnology*, *8*, 67–84.

12. Benoit-Norris, C., Cavan, D. A., & Norris, G. (2012). Identifying social impacts in product supply chains: Overview and application of the social hotspot database. *Sustainability*, *4*(9), 1946–1965.
13. Berne, R. W., & Schummer, J. (2005). Teaching societal and ethical implications of nanotechnology to engineering students through science fiction. *Bulletin of Science, Technology & Society*, *25*(6), 459–468.
14. Bernstam, M. S. (1986). Competitive human markets, interfamily transfers, and below-replacement fertility. *Population and Development Review*, *12*, 111–136.
15. Bhattacharyya, A., Datta, P. S., Chaudhuri, P., & Barik, B. R. (2011). Nanotechnology—A new frontier for food security in socio-economic development. In *Disaster Risk Vulnerability Conference* (Vol. 3, pp. 116–120).
16. Bordogna, C. M., & Albano, E. V. (2007). Statistical methods applied to the study of opinion formation models: A brief overview and results of a numerical study of a model based on the social impact theory. *Journal of Physics: Condensed Matter*, *19*(6), 065144.
17. Bornmann, L. (2013). What is societal impact of research and how can it be assessed? A literature survey. *Journal of the American Society for Information Science and Technology*, *64*(2), 217–233.
18. Bürgi, B. R., & Pradeep, T. (2006). Societal implications of nanoscience and nanotechnology in developing countries. *Current Science*, *7*, 645–658.
19. Chou, K. T., & Liou, H. M. (2011). Risk and ethical governance of nano-convergence technology: An initial comparison of the technological impact assessment between Korea and Taiwan. *Asian Journal of WTO & International Health Law & Policy*, *6*, 235.
20. Cohen, S., & Syme, S. L. (1985). Issues in the study and application of social support. *Social Support and Health*, *3*, 3–22.
21. Corley, E. A., Kim, Y., & Scheufele, D. A. (2016). Scientists' ethical obligations and social responsibility for nanotechnology research. *Science and Engineering Ethics*, *22*, 111–132.
22. De Jong, S., Barker, K., Cox, D., Sveinsdottir, T., & Van den Besselaar, P. (2014). Understanding societal impact through productive interactions: ICT research as a case. *Research Evaluation*, *23*(2), 89–102.
23. Doering, R. (2001). Societal implications of scaling to nanoelectronics. *Societal Implications of Nanoscience and Nanotechnology*, *8*(2), 68–75.

24. Downe, J., Cowell, R., & Morgan, K. (2016). What determines ethical behavior in public organizations: Is it rules or leadership? *Public Administration Review*, *76*(6), 898–909.
25. Eosco, G. M., Tallapragada, M., McComas, K. A., & Brady, M. (2014). Exploring societal and ethical views of nanotechnology REUs. *NanoEthics*, *8*, 91–99.
26. Fisher, E. (2007). The convergence of nanotechnology, policy, and ethics. *Advances in Computers*, *71*, 273–296.
27. Floridi, L. (2010). The ethics of the information society in a globalized world. In *The Cambridge Handbook of Information and Computer Ethics* (Vol. 5, pp. 271–283). Cambridge University Press.
28. Gajdzik, B., Grabowska, S., & Saniuk, S. (2021). Key socio-economic megatrends and trends in the context of the industry 4.0 framework. In *Forum Scientiae Oeconomia* (Vol. 9, pp. 5–22).
29. Gelfert, A. (2012). Nanotechnology as ideology: Towards a critical theory of 'converging technologies. *Science, Technology and Society*, *17*(1), 143–164.
30. Granitz, N. A. (2003). Individual, social, and organizational sources of sharing and variation in the ethical reasoning of managers. *Journal of Business Ethics*, *42*, 101–124.
31. Grunwald, A. (2020). Nanotechnology—A new field of ethical in□uiry? In *The Ethics of Nanotechnology, Geoengineering, and Clean Energy* (Vol. 2, pp. 17–31). Routledge.
32. Gupta, A., Deokar, A., Iyer, L., Sharda, R., & Schrader, D. (2018). Big data & analytics for societal impact: Recent research and trends. *Information Systems Frontiers*, *20*, 185–194.
33. Gupta, N., Fischer, A. R., George, S., & Frewer, L. J. (2013). Expert views on societal responses to different applications of nanotechnology: A comparative analysis of experts in countries with different economic and regulatory environments. *Journal of Nanoparticle Research*, *15*, 1–15.
34. Jung, I. (2009). Ethical judgments and behaviors: Applying a multidimensional ethics scale to measuring ICT ethics of college students. *Computers & Education*, *53*(3), 940–949.
35. Kalshoven, K., Den Hartog, D. N., & De Hoogh, A. H. (2011). Ethical leader behavior and big five factors of personality. *Journal of Business Ethics*, *100*, 349–366.
36. Kantor, J., & Weisberg, J. (2002). Ethical attitudes and ethical behavior: Are managers role models? *International Journal of Manpower*, *23*(8), 687–703.

37. Kim, J. M. (2011). The perspective of nanotechnology and its convergence with future information technology. *ECS Transactions*, *35*(3), 5.
38. Koepsell, D. (2011). *Innovation and Nanotechnology: Converging Technologies and the End of Intellectual Property*. Bloomsbury Academic.
39. Kump, B., & Scholz, M. (2022). Organizational routines as a source of ethical blindness. *Organization Theory*, *3*(1), 2631–2635.
40. Laurent, B. (2007). Diverging convergences: Competing meanings of nanotechnology and converging technologies in a local context. *Innovation*, *20*(4), 343–357.
41. Laurent, L., & Petit, J. C. (2005). Nanoscience and its convergence with other technologies. *HYLE–International Journal for Philosophy of Chemistry*, *11*(1), 45–76.
42. Lopez, J. (2003). Compiling the ethical, legal and social implications of nanotechnology. *Health Law Review*, *12*, 24.
43. Lu, L. Y., Lin, B. J., Liu, J. S., & Yu, C. Y. (2012). Ethics in nanotechnology: What's being done? What's missing? *Journal of Business Ethics*, *109*, 583–598.
44. Ma'Ayan, Y., & Carmeli, A. (2016). Internal audits as a source of ethical behavior, efficiency, and effectiveness in work units. *Journal of Business Ethics*, *137*, 347–363.
45. Magalhães, A., dos Santos, N. R., & Pais, L. (2019). Multi-source research designs on ethical leadership: A literature review. *Business and Society Review*, *124*(3), 345–364.
46. Masara, B., van der Poll, J. A., & Maaza, M. (2021). A nanotechnology-foresight perspective of South Africa. *Journal of Nanoparticle Research*, *23*, 1–22.
47. McComas, K. A. (2012). Researcher views about funding sources and conflicts of interest in nanotechnology. *Science and Engineering Ethics*, *18*, 699–717.
48. Mitchell, M. S., Reynolds, S. J., & Treviño, L. K. (2020). The study of behavioral ethics within organizations: A special issue introduction. *Personnel Psychology*, *73*(1), 5–17.
49. Molas-Gallart, J., & Tang, P. (2011). Tracing 'productive interactions' to identify social impacts: An example from the social sciences. *Research Evaluation*, *20*(3), 219–226.
50. Ponkratova, L. A., Feoktistov, S. V., Adashova, T. A., Nikolskaya, E. Y., & Livson, M. V. (2021). Tourism development prospects in the context of global socio-economic trends. *Journal of Environmental Management*

and Tourism, *11*(8), 1988–1993.

51. Prigulnyy, A. (2012). The most important directions of development of technology convergence and the basic scope of nanoproducts. *Social and Natural Sciences Journal*, *6*, 987–990.
52. Pulido, C. M., Ruiz-Eugenio, L., Redondo-Sama, G., & Villarejo-Carballido, B. (2020). A new application of social impact in social media for overcoming fake news in health. *International Journal of Environmental Research and Public Health*, *17*(7), 2430.
53. Roco, M. C. (2005). The emergence and policy implications of converging new technologies integrated from the nanoscale. *Journal of Nanoparticle Research*, *7*, 129–143.
54. Roco, M. C., & Bainbridge, W. S. (2007). Public perceptions. In *Nanotechnology: Societal Implications: Individual Perspectives* (Vol. 2, pp. 241–295). Dordrecht: Springer Netherlands.
55. Roiko, A., Mangoyana, R. B., McFallan, S., Carter, R. W., Oliver, J., & Smith, T. F. (2012). Socio-economic trends and climate change adaptation: The case of South East Queensland. *Australasian Journal of Environmental Management*, *19*(1), 35–50.
56. Ronteltap, A., Fischer, A. R., & Tobi, H. (2011). Societal response to nanotechnology: Converging technologies–converging societal response research? *Journal of Nanoparticle Research*, *13*, 4399–4410.
57. Rosenbaum, J. R., Bradley, E. H., Holmboe, E. S., Farrell, M. H., & Krumholz, H. M. (2004). Sources of ethical conflict in medical housestaff training: A qualitative study. *The American Journal of Medicine*, *116*(6), 402–407.
58. Sahoo, S. (2010). Socio-ethical issues and nanotechnology development: Perspectives from India. In *10th IEEE International Conference on Nanotechnology* (Vol. 3, pp. 1205–1210). IEEE.
59. Sarma, S. D., & Chaudhury, S. (2009). Socio-economic implications of nanotechnology applications: A case of copper and copper dependent countries. *Nanotechnology Law & Business*, *6*, 278.
60. Schummer, J. (2006). 'Societal and ethical implications of nanotechnology': Meanings, interest groups, and social dynamics. In *Nanotechnology Challenges: Implications for Philosophy, Ethics and Society* (Vol. 2, pp. 413–449).
61. Song, C. H., Elvers, D., & Leker, J. (2017). Anticipation of converging technology areas—A refined approach for the identification of attractive fields of innovation. *Technological Forecasting and Social Change*, *116*, 98–115.

62. Sparrow, R. (2020). The social impacts of nanotechnology: An ethical and political analysis. In *The Ethics of Nanotechnology, Geoengineering, and Clean Energy* (Vol. 3, pp. 175–185). Routledge.
63. Stevens, B. (1999). Communicating ethical values: A study of employee perceptions. *Journal of Business Ethics*, *20*, 113–120.
64. Suchman, M. C. (2001). Envisioning life on the nano-frontier. *Societal Implications of Nanoscience and Nanotechnology*, *5*(2), 211–216.
65. Sykes, G. M., & Matza, D. (2017). Techniques of neutralization: A theory of delinquency. In *Delinquency and Drift Revisited, Volume 21* (Vol. 1, pp. 33–41). Routledge.
66. Talala, K., Huurre, T., Aro, H., Martelin, T., & Prättälä, R. (2009). Trends in socio-economic differences in self-reported depression during the years 1979–2002 in Finland. *Social Psychiatry and Psychiatric Epidemiology*, *44*, 871–879.
67. Tegart, G. (2004). Nanotechnology: The technology for the twenty-first century. *Foresight*, *6*(6), 364–370.
68. Tenner, E. (2001). Nanotechnology and unintended consequences. *Societal Implications of Nanoscience and Nanotechnology*, *4*(3), 241–245.
69. Van Wart, M. (1996). The sources of ethical decision making for individuals in the public sector. *Public Administration Review*, *56*(3), 525–533.
70. Viseu, A., & Maguire, H. (2012). Integrating and enacting 'social and ethical issues' in nanotechnology practices. *NanoEthics*, *6*, 195–209.
71. Wetter, K. J. (2006). Implications of technologies converging at the nano-scale. In *Nanotechnologien Nachhaltig Gestalten. Konzepte und Praxis für eine Verantwortliche Entwicklung und Anwendung* (Vol. 9, pp. 31–41). Institut für Kirche und Gesellschaft: Iserlohn, Germany.
72. Wilhelm, M., Hutchins, M., Mars, C., & Benoit-Norris, C. (2015). An overview of social impacts and their corresponding improvement implications: A mobile phone case study. *Journal of Cleaner Production*, *102*, 302–315.
73. Yadav, S. K., Khan, Z. A., & Mishra, B. (2013). Impact of nanotechnology on socio-economic aspects: An overview. *Reviews in Nanoscience and Nanotechnology*, *2*(2), 127–142.
74. Yamin, M. A. Y. (2020). The relationship between RIGHT ethical behavior perspective, demographic factors, and BEST ethical performance. *International Review of Management and Marketing*, *10*(1), 27.

INDEX

D

E

F

G

H

O

P

Q

R

S

T

U

V

W

X

Z

FSC
www.fsc.org
MIX
Papier | Fördert gute Waldnutzung
FSC® C083411

Kolibri 360 GmbH
Ferdinand-Jühlke-Straße 7
99095 Erfurt, Deutschland
produktsicherheit@kolibri360.de